MECHANICAL ENGINEERING THEORY AND APPLICATIONS

RECENT TRENDS IN FATIGUE DESIGN

MECHANICAL ENGINEERING THEORY AND APPLICATIONS

Additional books in this series can be found on Nova's website under the Series tab.

Additional e-books in this series can be found on Nova's website under the e-book tab.

MECHANICAL ENGINEERING THEORY AND APPLICATIONS

RECENT TRENDS IN FATIGUE DESIGN

RICARDO BRANCO
EDITOR

nova publishers
New York

Library of Congress Cataloging-in-Publication Data

Recent trends in fatigue design / editor, Ricardo Branco (Polytechnic Institute of Coimbra, ISEC, Department of Mechanical Engineering, Coimbra, Portugal).
pages cm -- (Mechanical engineering theory and applications.)
Includes bibliographical references and index.
ISBN: 978-1-61668-410-5 (hardcover)
1. Materials--Fatigue. I. Branco, Ricardo (Mechanical engineer)
TA418.38.R43 2014
620.1'126--dc23
2014035654

Published by Nova Science Publishers, Inc. † New York

CONTENTS

PREFACE

Fatigue is progressive and localised structural damage that occurs when materials are subjected to cyclic loadings. Therefore, in order to produce durable products against fatigue and to prevent catastrophic failures it is essential to develop feasible fatigue design methods as well as to understand the fatigue phenomenon and its micro-mechanisms. This book covers leading-edge research on fundamental areas of fatigue. The aim is to inform those who might not to be at the forefront of recent developments and the wide range of applications for which solutions are now possible. The book is organised as follows:

- The first chapter (written by Edson Denner Leonel) applies reliability-based design optimisation approaches to fatigue crack growth problems;
- The second chapter (written by Aleksandar Grbović and Boško Rašuo) is focused on fatigue life prediction models for aircraft structural components;
- The third chapter (written by N. Ranganathan and R. Leroy) addresses the notch effect on fatigue crack initiation life considering constant- and variable-amplitude loading;
- The fourth chapter (written by Elias Merhy) deals with damage-tolerant design approaches for automotive cast alloy components experiencing thermo-mechanical fatigue loading;
- The fifth chapter (written by Fang Wang and Weicheng Cui) is focused on the recent developments of fatigue crack growth models and their applications;

- The sixth chapter (M.C. Marinelli and S. A. N. Hereñú) is devoted to short fatigue crack behaviour of duplex stainless steels under different thermal conditions;
- The seventh chapter (written be Anghel Cernescu) tackles the role of microstructure on fatigue crack growth rates;
- The last chapter (written by K. V. Arun, K. K. Jadhav and Virupaxi Bagodi) is concerned with the effect of crack configuration on strain energy release rate and tearing modulus in structural materials.

Dr. Ricardo Branco
Department of Mechanical Engineering
Polytechnic Institute of Coimbra, Portugal
Tel: +351 239 790 200
E-mail: rbranco@isec.pt

In: Recent Trends in Fatigue Design
Editor: Ricardo Branco
ISBN: 978-1-61668-410-5

Chapter 1

RELIABILITY-BASED DESIGN OPTIMIZATION APPROACH APPLIED TO FATIGUE CRACK GROWTH

Edson Denner Leonel*
University of São Paulo, School of Engineering of São Carlos Department of Structural Engineering, Brazil

ABSTRACT

This chapter presents a Reliability-Based Design Optimization (RBDO) approach which is applied to analysis of structures subjected to crack propagation due to fatigue phenomenon. The RBDO model is based on the coupling of a mechanical model and a reliability algorithm, in which both are connected into an optimization problem. The mechanical model is constructed using the Boundary Element Method (BEM). Crack propagation phenomenon is efficiently modelled by BEM, due to its mesh reduction aspects. The BEM model is based on the dual BEM formulation, in which singular and hyper-singular integral equations are adopted to the construction of the algebraic system of equations. The First Order Reliability Method is coupled to BEM model in order to assess the probabilities of structural failure. This type of coupling, known in literature as direct coupling, is efficient and accurate as the limit state function remains implicit and its derivatives are determined directly from

* Corresponding Author address. Avenida Trabalhador São Carlense, 400. CEP 13.566-590. São Carlos-SP Brazil. Email:edleonel@sc.usp.br.

the numerical mechanical responses. The optimization problem is formulated aiming to determine the minimal structural cost taking into account costs related to construction, inspection, maintenance and failure. All these costs are weighted by the probability of failure associated with each limit state considered. Moreover, constraints related to structural safety are also considered. The RBDO model is applied in three different optimization scenarios to illustrate the applicability and robustness of this approach.

Keywords: Boundary Element Method, RBDO model, Probabilistic Analysis, Fatigue Crack Growth

INTRODUCTION

Fracture mechanics theory, especially applied to fatigue problems, has been widely studied by the scientific community in recent years, because this theory is consistent to explain the failure of structures due to crack propagation. When fatigue phenomenon is considered, the structure loses its mechanical integrity along time [1] as cracks grow as a function of load cycles and, consequently, a long time. Due to the brittle characteristic of this type of failure, fatigue structural failure generally leads to catastrophic consequences, normally involving human life loss [2]. Therefore, the modelling of fatigue phenomenon and the prevision of structural failure is an important task into domain of engineering structures.

To model accurately the structural mechanical behaviour of complex engineering structures, including complex geometries and boundary conditions, numerical techniques are required. Therefore, considering fatigue crack propagation, numerical models are required as structural geometry and boundary conditions change at each crack length increment. In this regard, the Boundary Element Method (BEM) is recognized as an accurate and efficient numerical technique, which handles properly complex crack propagation problems [3,4]. Compared to domain mesh approaches, BEM is more efficient, especially in solving mixed mode crack propagation problems due to its accuracy in stress concentration modelling and mesh reduction features, as only the body's boundaries are discretised [5,6].

Many developments have been presented in literature concerning numerical approaches applied to fatigue crack propagation modelling. Most of these models focus only deterministic aspects of the mechanical structural behaviour. For instance, cohesive crack growth formulations, numerical models applied to multiple crack propagation simulation and fatigue analyses were successfully addressed in the scope of deterministic field [7-12]. However, several mechanical systems, among them systems subjected to fatigue, are affected by uncertainties, which lead to non-deterministic mechanical behaviour. In this regard, more realistic structural modelling is achieved only considering probabilistic approaches [13].

Fatigue crack propagation is an inherently random phenomenon, hence traditional, deterministic solution methods often fail to accurately predict the structural behaviour. The main parameters governing the crack growth process are the structural geometry, the initial crack configuration, material properties (especially those related to crack propagation rates) and the loading history. These parameters are inherently random and best modelled using probability theory. Random fatigue and crack propagation problems have been successfully addressed by several authors [14-16]. However, its modelling assumptions are sometimes quite restrictive, leading to solutions that are often inappropriate for practical engineering applications.

In structural mechanics, design optimization is the decision-making process that aims at finding the best set of design variables which minimizes some cost model while satisfying some performance and safety requirements [17,18]. Due to the inconsistency between these two objectives, the optimal solutions often lie on the boundaries of the admissible space. Thus, these solutions are rather sensitive to uncertainties either in the parameters or in the models themselves. Reliability-Based Design Optimization (RBDO) is a concept that accounts for uncertainty all along the optimization process. Basically, the deterministic performance model is wrapped into a more realistic probabilistic constraint which is referred to as the failure probability [19-21].

RBDO has arisen from practical engineering problems, in which the structural modelling is required to reach the realistic model. Unfortunately, most of structural design parameters that relate loads to resistances are inherently random. Therefore, RBDO makes the gap between structural design and real structure become smaller [22-24].

In the present investigation, a RBDO model is applied to analysis of fatigue crack growth considering uncertainties. The main goal of this model concerns the determination of minimal structural cost taking into account costs related to construction, inspection, maintenance and failure. All these costs are weighted by the probability of failure associated with each limit state considered. Moreover, constraints related to structural safety are also considered.

The mechanical model used in this chapter is based on the dual BEM formulation, in which singular (displacement integral equation) and hyper-singular (traction integral equation) integral equations are adopted. The displacement integral equations are written for collocation points along one crack surface whereas traction integral equations are used for collocation points along the opposite crack surface. This technique avoids singularities of the resulting algebraic system of equations, despite the fact that the collocation points on the opposite crack lips have the same coordinates.

The probabilistic approach is based on the direct coupling method [25]. Considering this approach, the limit state function derivatives are calculated directly, based on the numerical responses of the mechanical model. The limit state function remains implicit and defined by the mechanical model responses. The probability of failure is calculated using the First Order Reliability Method (FORM). The optimization problem is solved using the algorithm Sequential Quadratic Programming (SQP), which allows quadratic convergence of the goal function. The RBDO model is applied into three different optimization scenarios in order to illustrate the applicability and robustness of this approach.

Linear Elastic Fracture Mechanics and Fatigue Models

Stress intensity factors depend on the crack and structural geometry as well as the stress field at the crack tip. For complex structures and boundary conditions, these parameters are only properly calculated by numerical models. In this chapter, stress intensity factors are evaluated using the displacement correlation technique. For 2D structures, the stress intensity factors for modes I (opening) and II (sliding) are given according the following expressions, [26]:

$$K_I = \sqrt{\frac{2 \cdot \pi}{r}} \frac{\xi}{(\kappa + 1)} u_o \tag{1}$$

$$K_{II} = \sqrt{\frac{2 \cdot \pi}{r}} \frac{\xi}{(\kappa + 1)} u_s \tag{2}$$

where K_I and K_{II} are the stress intensity factors for modes I and II, respectively, u_o is the crack opening displacement, u_s is the crack sliding displacement, r is the distance between the crack tip and the computational point (i.e., mesh node) and ξ and κ are material properties. In order to calculate stress intensity factors at the crack tip, these variables are evaluated for four pairs of mesh nodes near the crack tip. Then, the stress intensity factors at the crack tip are obtained from a local extrapolation process. This process leads to accurate results as presented by [4,12].

There are many techniques available for extracting stress intensity factors. Among them, the displacement correlation technique was used in this chapter because it is very efficient and simple. The stress intensity factors are evaluated using the displacements calculated on the crack surfaces at each iteration. Therefore, extra operations are not required. The *J*-integral is another parameter that can be used to accurately evaluate stress intensity factors. However, for analysis of multi-fractured bodies and coalescence simulation, its numerical implementation is not trivial. In this case, the definition of evaluation points surrounding the crack tip must be carefully studied.

In mixed mode propagation, it is required the evaluation of the effective stress intensity factor K, and the crack propagation angle θ_p. For this purpose, the maximum circumferential stress criterion is used, [26]. According to this criterion, the cracks are assumed to grow in direction θ_p, which is perpendicular to the maximum circumferential stress at the crack tip.

$$K = K_I \cos^3\left(\frac{\theta_p}{2}\right) - 3K_{II}\cos^2\left(\frac{\theta_p}{2}\right)\sin\left(\frac{\theta_p}{2}\right) \tag{3}$$

$$\tan\left(\frac{\theta_p}{2}\right) = \frac{1}{4}\left[\frac{K_I}{K_{II}} \pm \sqrt{\left(\frac{K_I}{K_{II}}\right)^2 + 8}\right] \tag{4}$$

The equations presented above can be applied to fatigue crack growth simulation. The fatigue life prediction is a challenging problem in engineering design, inspection and maintenance. It is important to accurately estimate the life distribution of mechanical and structural components with respect to the number of load cycles. For several types of materials, the crack growth ratio can be modelled by using Paris' law [27], which is expressed as follows:

$$\frac{da}{dN} = C\Delta K^{n} \tag{5}$$

in which a represents the crack length, N is the number of loading cycles, C and n are material constants and ΔK is the stress intensity factor range.

For each load cycle, the effective stress intensity factor is evaluated at minimum and maximum load levels, namely $K_{\min}$ and $K_{\max}$, respectively. If the factor value is bigger than the material toughness, brittle failure is considered and the analysis is stopped. In this case failure happens. When $K_{\max} < K_{Ic}$, the factor increment ΔK is calculated and compared to the threshold ΔK_{th}. If $\Delta K \leq \Delta K_{th}$, no crack propagation is considered. Otherwise, the crack growth ratio is calculated by Paris' law, Eq. (5). The fatigue failure is defined by the crack growth ratio during the fracture process. As recommended by [28], when the crack growth ratio da/dN is bigger than 0.1 mm/cycle structural failure occurs (the remaining life is negligible) and the analysis is stopped. In this case, the crack propagation becomes instable and the fracture occurs.

BOUNDARY INTEGRAL EQUATIONS

BEM has been widely applied in several engineering fields such as contact problems, fatigue and fracture mechanics due to its high accuracy and robustness in modelling strong stress concentration (i.e., singular stresses and displacements fields). In order to develop the integral representation required by BEM, a two-dimensional homogeneous elastic domain Ω with a boundary Γ is considered. The equilibrium equation can be expressed in terms of displacements as follows:

$$u_{i,jj} + \frac{1}{1-2\upsilon} u_{j,ji} + \frac{b_i}{\mu} = 0 \tag{6}$$

in which μ is the shear modulus, υ is the Poisson's ratio, u_i are components of the displacement field and b_i represents the body forces. Using Betti's theorem, a singular integral representation for displacements can be obtained (with no body forces), as follows:

$$c_{lk}(f,c)u_k(f)+\int_{\Gamma} P^*_{lk}(f,c)u_k(c)\,d\Gamma=\int_{\Gamma} P_k(c)u^*_{lk}(f,c)\,d\Gamma \tag{7}$$

where P_k and u_k are tractions and displacements on the boundary, respectively, the free term c_{lk} is equal to $\delta_{lk}/2$ for smooth contours and P^*_{lk} and u^*_{lk} are the fundamental solutions for tractions and displacements, [29].

Equation (7) is sufficient to construct a system of algebraic equations to analyse 2D elastic domains. However, for solids containing cracks, the use of only this equation to assemble the final system of equations will lead to a singular matrix. It happens because both crack surfaces are located along the same geometrical path. Despite this observation, it is possible to use only the singular integral representation, Eq. (7), for crack simulation. In this case, a finite gap between two crack surfaces has to be considered as well as an accurate integral scheme to evaluate the integrals along the quasi-singular elements.

Several BEM formulations have been proposed in literature to properly handle crack problems [1,3-12]. The dual boundary element formulation (DBEM) is one of the most popular BEM formulations used to analyse random crack growth. In this formulation, the singular integral representation, Eq. (7), is used to determine the algebraic representation related to collocation points defined along one crack surface, whereas hyper-singular integral representation is used to obtain the algebraic representation related to collocation points located along the opposite crack surface. For all external boundaries, the singular representation applied to the boundary collocation points is sufficient to obtain the required algebraic relations.

The hyper-singular integral (or traction integral) representation at the boundary or crack surface collocation points can be obtained from Eq. (7). First, this equation, which is written for an internal collocation point, is differentiated to obtain the integral representation in terms of the strains. Then, using Hooke's law, the stress integral representation is achieved. Finally, the integral representation of stresses for a boundary collocation is determined by carrying out the relevant limits. The Cauchy formula is applied to obtain the traction representation as follows:

$$\frac{1}{2}P_j(f)+\eta_k\int_{\Gamma} S_{kij}(f,c)u_k(c)d\Gamma = \eta_k\int_{\Gamma} D_{kij}(f,c)P_k(c)d\Gamma \quad (8)$$

in which terms S_{kij} and D_{kij} contain the new kernels evaluated from P_{lk}^* and u_{lk}^*, [29, 30].

BEM Algebraic Equations

Algebraic equations are obtained for DBEM from Eqs. (7) and (8) after dividing the boundary and crack surfaces into elements over which displacements and tractions are approximated. For simplicity, linear elements are used to approximate tractions, displacements and the structural geometry. Then, a convenient number of collocation points are selected to obtain the number of algebraic equations required to solve the problem. Along the crack surfaces, geometrically identical elements are defined along opposite surfaces. Therefore, opposite nodes will be defined by the same coordinates to approximate displacements and tractions along the crack surface. Boundary collocation points are defined either at the element ends coincident with the nodes or along the element if displacement and traction discontinuities are recommended. Thus, based on the discretisation described above, the algebraic form of Eq. (7) can be obtained as it applies only to boundary collocation points. In this case the boundary and crack surface values are related as follows:

$$H_b^b U_b + H_b^f U_f = G_b^b P_b + G_b^f P_f \quad (9)$$

where U_b and U_f are displacements assigned to the boundary and crack surface nodes, respectively, P_b and P_f are the boundary and crack surface tractions, respectively, and H_b^b, H_b^f, G_b^b and G_b^f are the corresponding matrices that account for displacement and traction effects. The subscript b indicates the collocation point positioned on the boundary and the superscripts specify the boundary (b) or crack surface (f) values.

Equation (9) shows four crack values associated with each pair of opposite crack nodes, two displacements U_f and two tractions P_f. Therefore, four algebraic equations are required at each pair of opposite crack nodes and two equations must be written at each node. Due the singularity observed on the kernel of Eq. (8), continuity of displacements derivative at the collocations must be assured. Therefore, the crack surface collocations are always defined at a point along the element, which avoids placement at the element end, where the continuity could not be assured using the classical polynomial approximation. For each pair of collocation points, Eq. (7) is used to determine two algebraic equations associated with one collocation point, whereas Eq. (8) is used to determine the other two algebraic equations associated with the opposite collocation point. Then, by combining these algebraic relations, the following set of algebraic equations is obtained:

$$H_f^b U_b + H_f^f U_f = G_f^b P_b + G_f^f P_f \tag{10}$$

where the subscript f in the matrices H_f^x and G_f^x indicates the equation for collocation along the crack line.

To evaluate the relevant integrals along elements, analytical expressions are used to perform integrals along singular elements. For non-singular elements, both analytical expressions and Gauss-Legendre integration scheme, coupled to sub-element process, are used [4,5]. Analytical expressions were used to integrate hyper-singular terms and numerical scheme to perform the relevant integrals in Eq. (7). Based on these procedures, Eq. (7) and Eq. (8) are transformed into algebraic representations with very low integration error.

Equations (9) and (10) can be used to obtain the system of algebraic equations for analysis of solids containing cracks. However, Eq. (9) and Eq. (10) should be modified to consider independent values defined along the left and right crack surfaces indicated by the subscripts r and ℓ, respectively:

$$H_b^b U_b + H_b^r U_r + H_b^\ell U_\ell = G_b^b P_b + G_b^r P_r + G_b^\ell P_\ell \tag{11}$$

$$H_f^b U_b + H_f^r U_r + H_f^\ell U_\ell = G_f^b P_b + G_f^r P_f^r + G_f^\ell P_\ell \tag{12}$$

Therefore, the vector U_f is now split into U_r and U_ℓ, and P_f becomes P_r and P_ℓ. The displacements and tractions on the crack surface are described according the normal, *n*, and parallel, *s*, directions to crack surfaces, as presented in Figure 1.

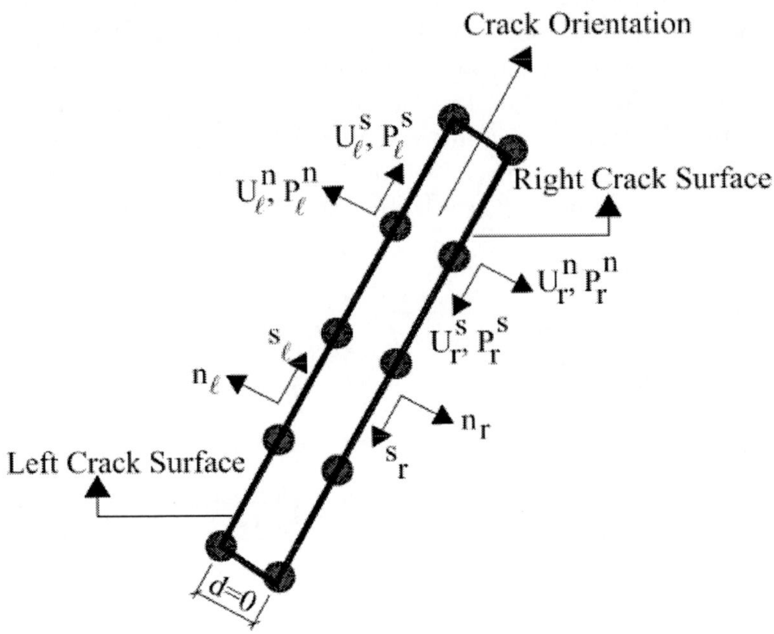

Figure 1. Right and left surface variables (n and s indicate normal and parallel directions to crack surface, respectively).

Equations (11) and (12) can be further modified by introducing the crack relative displacements u as follows:

$$U_\ell = u - U_r \tag{13}$$

The vector *u* contains the crack opening displacements, u_0, and the crack sliding displacement u_s. Crack opening displacement is defined as the relative displacements along the normal crack surface direction, whereas crack sliding displacement contains the relative displacements along the parallel crack surface direction. These values are evaluated for each pair of nodes positioned at the crack surfaces.

After these modifications, Eq. (11) and Eq. (12) become:

$$H_b^b U_b + \left(H_b^r - H_b^\ell\right) U_r + H_b^\ell u = G_b^b P_b + G_b^r P_r + G_b^\ell P_\ell \tag{14}$$

$$H_f^b U_b + \left(H_f^r - H_f^\ell\right) U_r + H_f^\ell u = G_f^b P_b + G_f^r P_r + G_f^\ell P_\ell \tag{15}$$

Equations (14) and (15) represent the equilibrium of the solid containing cracks. Assuming that along cracks surfaces the displacements U_r and the crack opening displacements u are unknowns and the tractions are always prescribed, Eq. (14) and Eq. (15) can be combined. The boundary conditions can first be applied by moving all of the boundary unknowns to the left-hand side of Eq. (14). As a result, A_b^b will be the block matrix that corresponds to the boundary displacement and tractions unknowns, which ones are stored in the vector X_b . Similarly, A_f^b refers the unknowns for the algebraic equations written at the crack surface collocation points. The right-hand side of both equations, which now contains only known values, is represented as the independent vector $\left\{F_b \;\; F_f\right\}^T$.

Therefore, the combination of Eq. (14) and Eq. (15) results in:

$$\begin{bmatrix} A_b^b & \left(H_b^r - H_b^\ell\right) & H_b^\ell \\ A_f^b & \left(H_f^r - H_f^\ell\right) & H_f^\ell \end{bmatrix} \begin{Bmatrix} X_b \\ U_r \\ u \end{Bmatrix} = \begin{Bmatrix} F_b \\ F_f \end{Bmatrix} \tag{16}$$

Based on the responses of Eq. (16), stress intensity factors are directly computed by Eq. (1), Eq. (2) and Eq. (3). Although linear, the above system of algebraic equations has to be solved by applying the loads in increments in order to trace the evolution of crack growth path. During the loading process, the direction and the length advance of each crack must be evaluated. The proposed mechanical BEM model is capable to simulate multiple crack distribution and coalescence conditions.

Considering fatigue crack propagation analysis, Eq. (16) has to be solved for maximum and minimum applied loads. The stress intensity factors are evaluated for each case, and the stress intensity factor range is calculated. Then, the structural life is predicted using Eq. (5).

RELIABILITY ANALYSIS

The reliability analysis aims at computing the probability of failure P_f regarding a specific failure scenario, known as limit state (note that reliability and failure probability are complementary).

The leading step in reliability assessment is to identify the basic set of random variables $X = \left[x_1, x_2, \ldots, x_n\right]^T$ for which uncertainties have to be considered. For all these variables, probability distributions are attributed to model randomness. These probability distributions are defined by physical observations, statistical studies, laboratory analysis and expert opinion. The number of random variables is an important parameter to determine the computing time consumed during the reliability analysis. In order to reduce the size of the random variable space, it is strongly recommended to consider as deterministic all variables which its uncertainties lead to minor effects on the value of the probability of failure.

The second step consists in defining a number of potentially critical failure modes. For each of them, a limit state function $G(X)$ separates the space into two regions: the safe domain, where $G(X) > 0$, and the failure domain where $G(X) < 0$. The boundary between these two domains is defined by $G(X) = 0$, known as the limit state itself. It is worth to mention that, in complex engineering analysis, an explicit expression for the limit state function is not often available. However, the limit state function can be calculated at some desired points by running the BEM analysis.

The failure probability is evaluated by the following integral, [13]:

$$P_f = \int_{G \leq 0} f_X\left(x_1, x_2, \ldots, x_n\right)\; dx_1, dx_2, \cdots dx_n \tag{17}$$

where $f_X\left(x_1, x_2, \ldots, x_n\right)$ is the joint density function of the variables X. As the evaluation of the above integration is impossible in practical applications, alternative procedures have been developed on the basis of the concept of reliability index β [31]. It is defined by the distance between the mean point and the failure domain in the normalized space of random variables. The reliability index allows the evaluation of the probability of failure, using the first order reliability method (FORM), as: $P_f = \Phi(-\beta)$, where $\Phi(\cdot)$ is the standard Gaussian cumulated distribution function.

When numerical mechanical methods are involved, the structural reliability analysis may be assessed by coupling directly the numerical mechanical model considered and FORM algorithm. This type of coupling is known in literature as direct coupling approach and will be discussed in the following section.

COUPLED MECHANICAL AND RELIABILITY APPROACH. DIRECT COUPLING METHOD

To consider probabilistic fatigue crack propagation, it is required to couple the reliability procedure with BEM model. In this chapter, this coupling is performed considering the direct method approach.

In this type of coupling, the basic procedure consists in directly coupling the reliability model, FORM, to the numerical mechanical model based on BEM. As described in the previous section, the limit state function defines the interface between safe and failure domains. For general engineering applications, this function can be expressed in terms of limit loads. For linear elastic fracture mechanics analysis, considering probabilistic approach, the limit state function is defined as follows:

$$G(X) = P_{\text{Resistant}}(X) - P_{\text{Applied}}(X) \tag{18}$$

in which $P_{\text{Resistant}}(X)$ is the load which leads to structural failure (regarding a specific failure scenario) and $P_{\text{Applied}}(X)$ is the applied external load.

In order to simulate probabilistic fatigue crack growth, the limit state function can be written in terms of the number of load cycles:

$$G(X) = N_{\text{Resistant}}(X) - N_{\text{Applied}}(X) \tag{19}$$

in which $N_{\text{Resistant}}(X)$ is the number of load cycles corresponding to structural failure and $N_{\text{Applied}}(X)$ is the number of load cycles applied along the structural service life.

To give invariance measure of safety, the random variables, defined in the physical space, are transformed into independent standard Gaussian variables [31], by using appropriate probabilistic transformation. Figure 2 illustrates this transformation where the limit state function $G(X)$, defined in the physical space, is transformed into $H(U)$ in the standard normalized space. In this transformation, $U = [u_1, u_2, \ldots, u_n]^T$ denotes the standard Gaussian variables.

The reliability index β is given by the minimum distance between a point on the limit state curve and the origin of standard space. In order to determine this point and calculate the probability of failure, a constrained minimization procedure has to be performed. Therefore, the optimization problem for β index determination can be stated as follows:

$$\begin{aligned} &\text{find: } U^* \\ &\text{which minimizes: } \beta = \sqrt{U^T \cdot U} \\ &\text{subject to: } \mathrm{H}(U) = 0 \end{aligned} \tag{20}$$

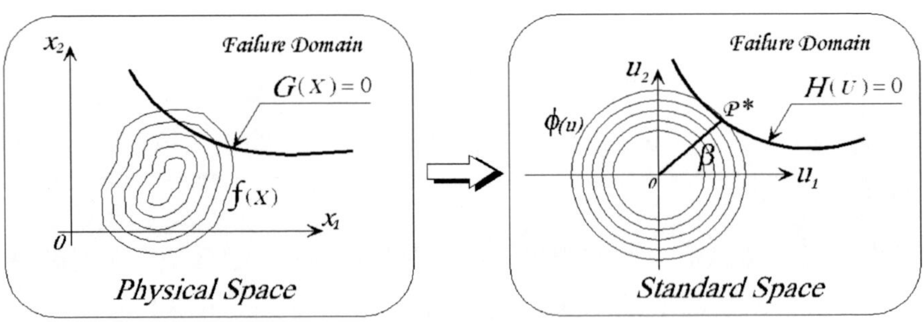

Figure 2. Isoprobabilistic transformation from physical to standard space.

The solution of the above constrained minimization problem converges to the failure point nearest to the standard normalized space origin. This point is known in literature as *design point* or the most probable failure point X*. The reliability index β can be achieved by applying the Rackwitz-Fiessler algorithm [32] directly to the mechanical model.

As the structural resistance (failure structural load or number of resisting load cycles) is known point by point using BEM model, the structural resistance is implicit and a function of the BEM model. In this case, the limit state function values and its derivatives are evaluated directly by running the BEM analysis. For the limit state function derivatives evaluation a numerical procedure is adopted. In this work, the forward finite difference scheme was chosen. The numerical error due to the finite difference may affect the convergence of the coupled procedure, as well as its accuracy. However, for all problems studied in this chapter, accurate and stable convergence ratio was observed.

It is worth to mention that accurate and stable mechanical models, as one used in this work, do not introduce approximations on the probabilistic analysis by adopting this approach. It happens because the probabilistic behaviour, limit state function values and derivatives, are evaluated directly by using the mechanical responses determined by the numerical model. When the structural resistance is accurately evaluated, the probabilistic analysis is accurately performed. Moreover, stable mechanical models lead to smooth and faster convergence of FORM, which become this approach efficient and faster when compared against other reliability algorithms such as Monte Carlo simulation and response surface method, for instance.

It is important to mention that the coupled model involving mechanical and direct coupling approaches was already validated in previous works of the author as presented in [2,25].

RBDO Model

In this section, the Reliability-Based Design Optimization (RBDO) model proposed is applied to analysis of fatigue crack growth. The goal of this model addresses the minimization of the structural volume, which reflects the minimization of the structural material cost, regarding some particular scenarios taking into account constraints related to structural safety. The scenarios proposed consider a combination among construction, inspection, maintenance and failure costs. Therefore, this model allows the determination of the structural volume that leads to the minimal cost regarding construction (structural material), inspection and maintenance procedures as well as failure considering a probability of failure target.

It is important to mention that the analyses performed by RBDO model accounts two different domains. The first domain is related to reliability, in which random variables and probabilistic features are considered. The second domain concerns the physical domain, in which variables related to structural geometry are calculated based on optimization procedures and safety constraints.

Safety constraints are evaluated at each iteration of the optimization problem and its values are determined based on the reliability model. The dependence between the probability of failure and the variables considered on optimization problem (structural dimensions, for instance) is expressed by a complete quadratic polynomial. The constants of this quadratic polynomial are obtained based on a simple least square procedure, which is performed considering the probability of failure values assessed by the reliability model considering the input values given by optimization algorithm.

Therefore, the RBDO problem is formulated based on the following equation:

$$\begin{aligned} &\textit{Minimize} && f(y) \\ &\textit{Subjected to} && \beta_{\mathrm{Target}} - \beta_{Structural}(x,y,z) \leq 0 \\ & && h_i(y) \leq 0 \end{aligned} \quad (21)$$

in which x indicates the random variables considered in the reliability problem, y contains the variables considered in the optimization problem and z represents the variables related to mechanical model, $f(y)$ is the goal function to be minimized, which contains structural dimensions and the costs related to construction, inspection, maintenance and failure, β_{Target} represents the reliability index target chosen by the analyst, $\beta_{\mathrm{Structural}}(x,y,z)$ is the complete polynomial expression that relates the reliability index to variables to be optimized and $h_i(y) \leq 0$ contains the constraints as maximum and minimum structural dimensions, for instance. It is worth to mention that the variables to be optimized, y, and the random variables, x, can be the same. The optimization problem proposed in Eq.(21) is solved consistently and faster using the optimization algorithm Sequential Quadratic Programming (SQP), as presented in [33].

The safety level target assumed in the analysis is considered by β_{Target}, which depends on the consequences of failure [34]. Onoufriou and Frangopol proposed some values of reliability index target and its correspondence with the consequence of failure. These values are presented in Table 1 and it was adopted in the analyses performed in this work.

Table 1. Reliability index target given by [34]

Consequences of Failure	Reliability Index Target	Probability of Failure
Serious	3.7	1.1E-04
Not Serious	3.1	9.7E-03
Local Effects	2.3	1.0E-02
Not important	1.0	1.0E-01

APPLICATIONS

Three applications involving the RBDO model proposed in this chapter are presented in the sequence of this text. This model is applied to the determination of: minimal structural volume regarding a reliability index target, the time (number of load cycles) for the start of maintenance procedures taking into account a structural safety target and both parameters already mentioned considering associated costs and a structural safety level target. These scenarios are modelled only by changing the goal function on the optimization problem and the considered constraints.

Determination of Time for Maintenance Procedures

This application aims at determining the time instant in which inspection and maintenance procedures have to be performed in order to keep the structural safety at a desirable level. The time instant is a function of the number of applied load cycles. Therefore, the RBDO problem to be solved is defined by as follows:

Minimize $$f(x,y,z) = \left|\beta_{Structural}(x,y,z) - \beta_{\text{Target}}\right| \quad (22)$$

It is important to mention that in this application constraints related to structural geometry are not considered. The goal of this analysis concerns the determination of the number of load cycles, which depends on time, that leads to the structural reliability index equal to the one predefined by the analyst, i.e., the reliability index target.

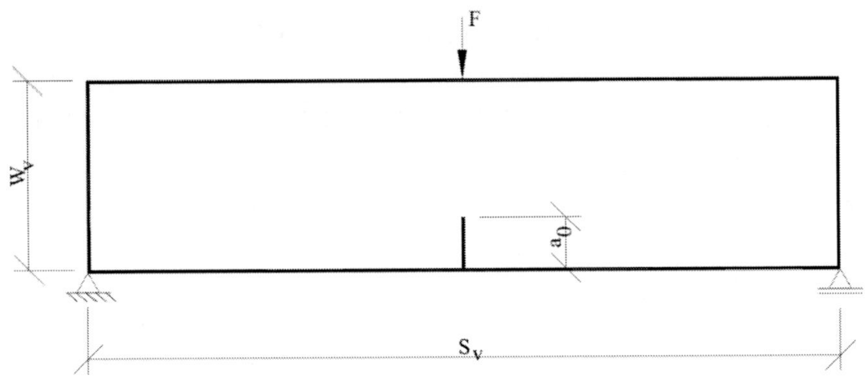

Figure 3. Structure analyzed in this chapter.

Table 2. Deterministic and random variables considered in this application

Variable	Distribution	Mean	Std Deviation
Young's modulus *E(GPa)*	deterministic	210	-
Poisson's ratio *v*	deterministic	0.20	-
Material Toughness K_c $(MN/m^{1.5})$	deterministic	104	-
Length S_v *(m)*	deterministic	5.0	-
High W_v *(m)*	deterministic	1.25	-
Paris Law parameter *n*	deterministic	2.70	-
Paris coefficient *C* $(m^{1-1.5n}/cycle/kN^n)$	lognormal	$3.0 \cdot 10^{-10}$	$1.8 \cdot 10^{-10}$
Applied load *F(kN)*	normal	5.0	0.80
Initial crack notch a_0 *(m)*	normal	0.01	0.003

The structure considered in this analysis is presented in Figure 3. It is a three point bended beam containing an initial notch at its middle span. The applied load is cyclic, which presents a complete load and unload behaviour at each load cycle. The crack grows as the cyclic load is applied. Then, fatigue effects are accounted and modelled by BEM.

The parameters values considered in this application are presented in Table 2. The reliability index target chosen is equal to 3.10. This value is obtained based on data presented in Table 1.

Considering the random parameters presented in Table 2, the number of load cycles that becomes null Eq. (22) is determined. The evolution of the structural reliability index as a function of the number of applied load cycles is presented in Figure 4. This analysis was carried out using 26 runs of the reliability model, which lead to 104 mechanical BEM model runs.

According to this figure, the structural reliability index reaches the one target at each $3.942 \cdot 10^5$ load cycles. Therefore, inspection and maintenance procedures should be performed in this time instant in order to keep the structural safety at the desired level given by β_{Target}. If perfect maintenance procedures are assumed, the structure recovers its initial safety level just after maintenance. Therefore, the evolution of the structural reliability index along time (number of load cycles) has the behaviour presented in Figure 4 as long as perfect maintenance is assumed.

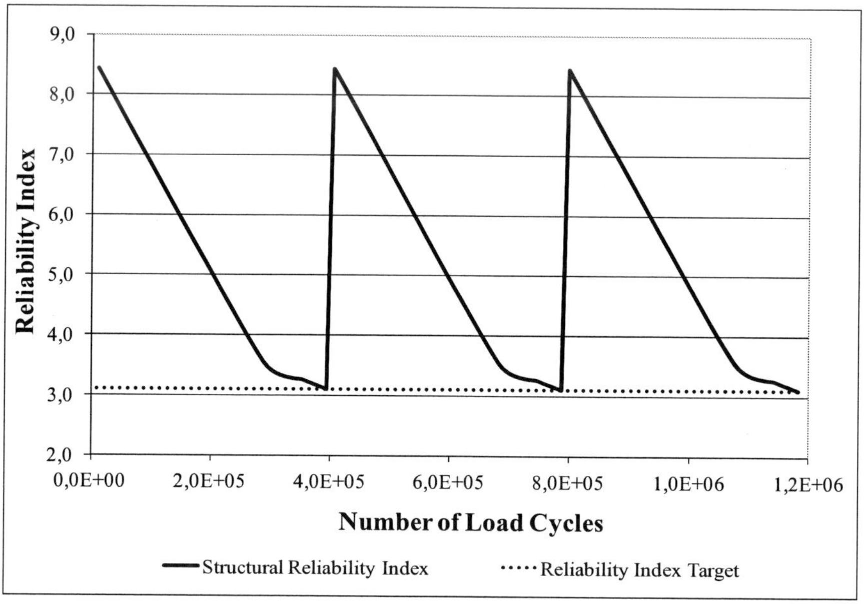

Figure 4. Evolution of the structural reliability index along time.

Determination of Structural Dimensions Considering a Safety Level Target

The second application of this chapter applies the proposed RBDO model for the determination of structural dimensions considering constraints related to structural safety. The structure considered in this analysis was presented in Figure 3, which is a three point bended beam. The load was assumed as cyclic with a complete load and unload behaviour at each load cycle. The structural collapse is modelled by BEM assuming the presence of fatigue effects. Therefore, the structural failure involves the crack propagation due to fatigue.

The parameters values considered in this application are presented in Table 3. The reliability index target chosen is equal to 2.30. This value is obtained based on data presented in Table 1.

The goal of this analysis concerns the determination of the structural thickness value, T_v, that leads to the minimal structural volume taking into account the reliability index target. Then, a structural geometry parameter is determined based on a constraint related to the desirable structural safety level.

Table 3. Deterministic and random variables considered in this application

Variable	Distribution	Mean	Std Deviation
Young's modulus *E(GPa)*	deterministic	210	-
Poisson's ratio v	deterministic	0.30	-
Material Toughness K_c $(MN/m^{1.5})$	deterministic	104	-
Length S_v *(m)*	deterministic	5.0	-
High W_v *(m)*	deterministic	1.25	-
Paris Law parameter n	deterministic	2.70	-
Paris coefficient C $(m^{1-1.5n}/cycle/kN^{n})$	lognormal	$3.0 \cdot 10^{-10}$	$1.8 \cdot 10^{-10}$
Applied load *F(kN)*	normal	5.0	0.80
Initial crack notch a_0 *(m)*	normal	0.01	0.003
Applied load cycles (*cycles*)	deterministic	$2.5 \cdot 10^4$	-

Therefore, the optimization problem to be solved is presented as follows:

$$\begin{aligned} &\text{Minimize} && f(y) = S_v W_v T_v \\ &\text{Subjected to} && \beta_{\text{Target}} - \beta_{Structural}(x, y, z) \leq 0 \end{aligned} \tag{23}$$

It is important to stress that beam's length and high were kept constant and deterministic during optimization and probabilistic iterations. This analysis was carried out using 18 runs of the reliability model which required 72 runs of the mechanical BEM model.

The evolution of the value of structural thickness along the iterations of the optimization problem as well as its numerical convergence behaviour are presented in Figure 5. According to this figure, four iterations of the optimization algorithm were required in order to reach the convergence of the RBDO model.

The final value of structural thickness obtained is 0.2965 m which leads to structural volume equal to 1.8533 m^3. It is important to inform that for the convergence configuration reached, the structural reliability index achieved is equal to the target, indicating that this constraint is active.

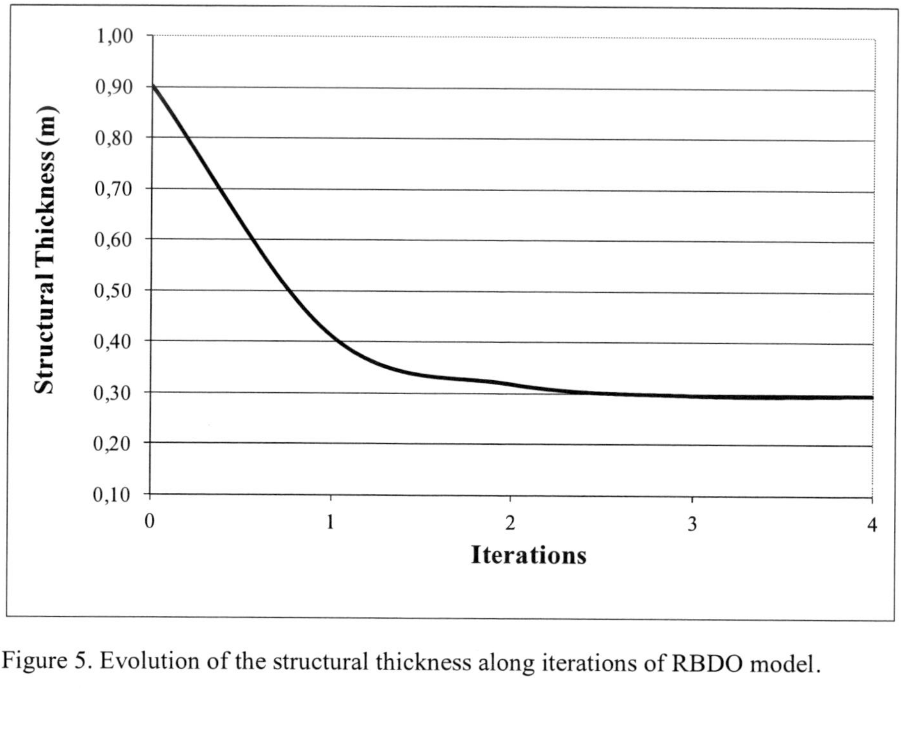

Figure 5. Evolution of the structural thickness along iterations of RBDO model.

In order to improve the study of the RBDO problem considered, a new RBDO analysis was also carried out considering that both structural thickness and high are determined assuming constraints based on structural safety. In this analysis, the random and deterministic parameters are the same presented in Table 3. The difference concerns the optimization problem, which is described by the following conditions:

$$\begin{aligned} &\textit{Minimize} && f(y) = S_v W_v T_v \\ &\textit{Subjected to} && \beta_{\text{Target}} - \beta_{Structural}(x, y, z) \leq 0 \\ & && 0.30 - T_v \leq 0 \end{aligned} \tag{24}$$

This analysis required 6 reliability model runs resulting into 24 mechanical BEM model runs. The convergence history for beam's thickness and high is presented in Figure 6.

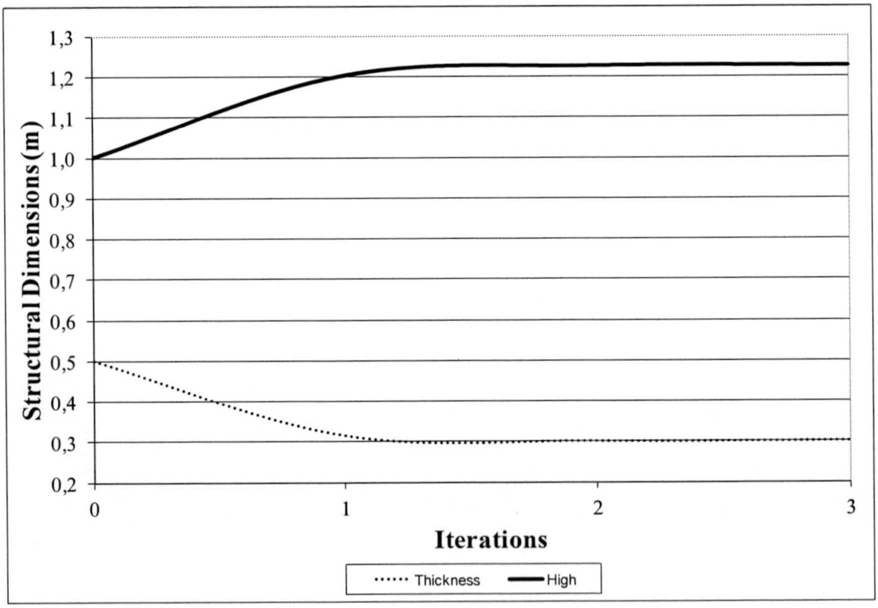

Figure 6. Evolution of the structural thickness and high along iterations of RBDO model.

According to this figure, faster convergence of RBDO model is observed. On the second iteration of the optimization problem good results were already observed. The values achieved at the end of the numerical procedure are equal to: beam's thickness 0.300 m and beam's high 1.226 m. These geometrical values result into structural volume of 1.8390 m^3. It is worth to inform that both constraints presented in Eq. (24) were active at the end of the iterative process, indicating the accuracy of the optimization model.

Determination of Structural Dimensions and Time for Inspection and Maintenance Procedures Considering Costs

The last application considered in this chapter addresses the analysis of optimization based on reliability constraints of the structure presented in Figure 3. It is a three point bended beam with an initial notch at its middle span. The goals of this analysis are the determination of the structural thickness and the time for inspection and maintenance procedures taking into account the cost of all structural life and the reliability index target.

The structure was analysed considering fatigue effects, which limit states are modelled by BEM. Then, the load was assumed as cyclic with a complete load and unload behaviour at each load cycle. Consequently, the crack grows along load cycles.

The optimization problem to be solved is defined by the following conditions:

$$\begin{array}{ll} \textit{Minimize} & S_v W_v T_v C_{Constr} + C_{Failure} P_{Failure} + C_{Inspec}\left(1 - P_{Failure}\right) + C_{\mathrm{Maint}}\left(1 - P_{Failure}\right) \\ \textit{Subjected to} & \beta_{\mathrm{Target}} - \beta_{Structural}\left(x, y, z\right) \le 0 \\ & \beta_{\mathrm{Target\ Inspection}} - \beta_{Structural}\left(x, y, z\right) \le 0 \\ & 0.30 - T_v \le 0 \end{array} \quad (25)$$

in which S_v, W_v and T_v are the structural dimensions, $C_{Failure}$ and $P_{Failure}$ represent the cost of structural failure and the probability of failure, respectively, C_{Inspec} and C_{Maint} indicate the costs related to inspection and maintenance procedures, respectively. β_{Target} and $\beta_{\mathrm{Target\ Inspection}}$ reflect the structural safety level desired by the analyst regarding the scenarios of structural failure and structural failure at inspection time, respectively.

In order to avoid the influence of the monetary unit adopted, the costs presented in Eq.(25) are considered as a ratio of the cost due to the structural failure. In the present analysis, the ratio costs presented in Table 4 were considered. The numerical analysis was carried out considering the parameters presented in Table 5.

Table 4. Ratio costs assumed in this application

$C_{\text{Constr}}/C_{\text{Failure}}$	$C_{\text{Maint}}/C_{\text{Failure}}$	$C_{\text{Inspec}}/C_{\text{Failure}}$	$C_{\text{Failure}}/C_{\text{Failure}}$
0.15	0.25	0.05	1.00

The reliability indexes target chosen for both global failure and maintenance failure are equal to 2.30. This value is obtained based on the data presented in Table 1.

As previously mentioned, in this analysis the structural thickness is determined considering the costs of construction, inspection, maintenance and structural failure as well as the probabilities of failure associated with each one. Moreover, this parameter is determined considering a safety level target defined by β_{Target} and $\beta_{\text{Target Inspection}}$. In addition to that, the number of load cycles (time instant) for maintenance procedures is also determined considering a predefined safety level given by $\beta_{\text{Target Inspection}}$. Therefore, this analysis aims at determining a structural geometric parameter and the time for inspection and maintenance procedures taking into account associated costs and a desired structural safety level.

Table 5. Deterministic and random variables considered in this application

Variable	Distribution	Mean	Std Deviation
Young's modulus *E(GPa)*	deterministic	210	-
Poisson's ratio v	deterministic	0.30	-
Material Toughness K_c $(MN/m^{1.5})$	deterministic	104	-
Length S_v (m)	deterministic	5.0	-
High W_v (m)	deterministic	1.25	-
Paris Law parameter n	deterministic	2.70	-
Paris coefficient C $(m^{1-1.5n}/cycle/kN^n)$	lognormal	$3.0 \cdot 10^{-10}$	$1.8 \cdot 10^{-10}$
Applied load $F(kN)$	normal	5.0	0.80
Initial crack notch a_0 (m)	normal	0.01	0.003

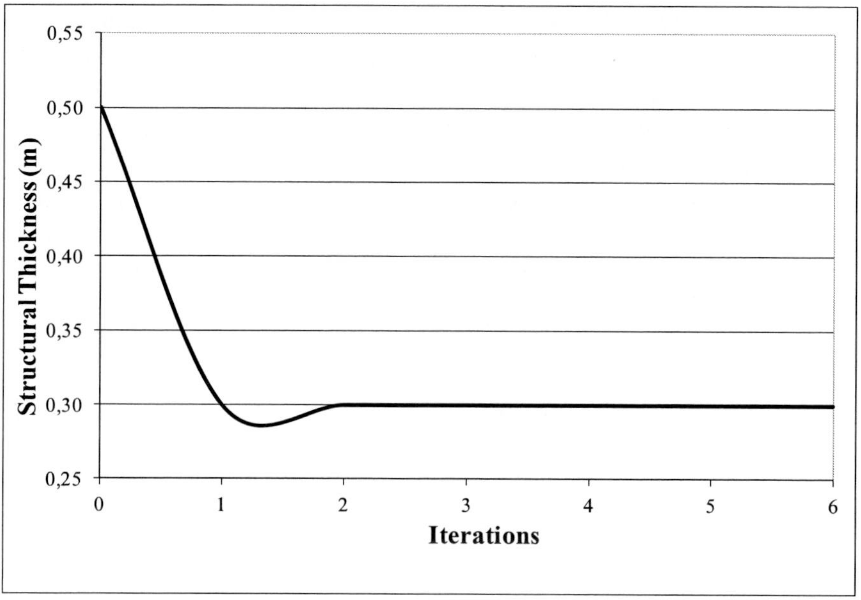

Figure 7. Evolution of the structural thickness along iterations of optimization problem.

Figure 8. Evolution of the load cycles for the determination of time instant of maintenance and inspection procedures.

The convergence histories for the two variables goal of this analysis are presented in Figure (7) and Figure (8). As illustrated in this figures, the RBDO model required 6 iterations to achieve the numerical convergence. It required 6 reliability model runs and, consequently, 36 mechanical BEM model runs.

According to Figures 7 and 8, the structural thickness achieved is equal to 0.30 m. It allows concluding that the geometrical constraint is active during the optimization iterative process. The time interval for maintenance and inspection procedures determined by the RBDO model proposed is equal to 25799 load cycles. Obviously, if the ratios presented in Table 4 be changed, the results presented in this paragraph will also change. However, these ratios are an input parameter in the model and it must be analysed according to the economic conditions of the country in which the structure is located.

CONCLUSION

This chapter addressed the analysis of optimization problems in engineering structures field considering constraints related to structural safety. The mechanical problem analysed involves the crack growth in two-dimensional structures due to fatigue phenomenon. In this mechanical problem, the crack grows as the load cycles are applied on the body, i.e., a long time. The fatigue crack growth problem is inherently random and best modelled considering the uncertainties under the problem variables.

The mechanical model developed to simulate the fatigue crack growth is based on the dual BEM formulation, which is efficient and accurate in modelling random crack growth. This BEM approach allows a simple body's discretisation as only the boundary is discretised. When crack growth is modelled only the new boundaries of cracks that had its length increased are remeshed. Therefore, the computational time consuming is considerable reduced as remeshing process is simple as well as the update process of algebraic matrices involved in BEM.

The probabilistic approach considered in this chapter has been successfully applied in literature. Direct coupling method is more efficient than simple response surface approach, as already shown by the author in [2,25]. The probabilities of failure are accurately determined by direct method according to the values of optimized variables considered in the optimization problem.

The optimization based on reliability was applied on three problems in this chapter. The minimization of structural volume, the determination of the number of load cycles for maintenance procedures and the coupling of both problems considering associated costs and a desirable safety level were accounted in the optimization problem. All these scenarios were consistently analysed by the proposed model, indicating its direct application into practical industrial problems.

It is worth to mention that the modelling of optimization scenarios different from those analysed in this chapter are possible. In this case, the analyst has to define the goal function, which must be based on safety constraints, taking into account the goal variables of his interest. Therefore, the proposed model is robust and its accuracy was already proved based on results presented.

Acknowledgments

Sponsorship of this research project by the São Paulo State Foundation for Research – FAPESP - is greatly appreciated. This research is a part of the activities scheduled by the research project USP/COFECUB 2012.1.672.1.0.

References

[1] Cisilino, A.P; Aliabadi, M.H. (1999). Three-dimensional boundary element analysis of fatigue crack growth in linear and non-linear fracture problems, *Engineering Fracture Mechanics*, 63, 713-733.

[2] Leonel, E.D; Beck, A.T; Venturini, W.S. (2011). On the performance of response surface and direct coupling approaches in solution of random crack propagation problems. *Structural Safety*. 33, 261-274.

[3] Dell'erba, D.N; Aliabadi, M.H. (2000). Three-dimensional thermo-mechanical fatigue crack growth using BEM, *International Journal of Fatigue*, 22, 261-273.

[4] Leonel, E.D; Venturini, W.S. (2010). Dual boundary element formulation applied to analysis of multi-fractured domains, *Engineering Analysis with Boundary Elements*, 34, 1092-1099.

[5] Leonel, E.D; Venturini, W.S; Chateauneuf, A. (2011). A BEM model applied to failure analysis of multi-fractured structures. *Engineering Failure Analysis*, 18, (6) 1538-1549.

[6] Yan, X. (2006). A finite main crack interaction with an arbitrarily oriented microdefect. *Engineering Failure Analysis*, 13, 971–980.

[7] Bouchard, P.O; Bay, F; Chastel, Y. (2002). Numerical modelling of crack propagation: automatic remeshing and comparison of different criteria. *Computer Methods in Applied Mechanics and Engineering*. 192, 3887–3908.

[8] Yan, X. (2007). Automated simulation of fatigue crack propagation for two-dimensional linear elastic fracture mechanics problems by boundary element method. *Engineering Fracture Mechanics*, 74, 2225-2246.

[9] Leonel, E.D; Venturini, W.S. (2010). Non-linear boundary element formulation with tangent operator to analyse crack propagation in quasi-brittle materials. *Engineering Analysis with Boundary Elements*. 34, 122–129.

[10] Kebir, H; Roelandt, J.M; Chambon, L. (2006). Dual boundary element method modelling of aircraft structural joints with multiple site damage. *Engineering Fracture Mechanics*. 73, 418-434.

[11] Romlay, F.R.M; Ouyang, H; Ariffin, A.K; Mohamed, N.A.N. (2010). Modeling of fatigue crack propagation using dual boundary element method and Gaussian Monte Carlo method, *Engineering Analysis with Boundary Elements*, 34, 297-305.

[12] Leonel, E.D; Venturini, W.S. (2011). Multiple random crack propagation using a boundary element formulation, *Engineering Fracture Mechanics*, 78, 1077–1090.

[13] Ditlevsen, O; Madsen, H.O. (1996). Structural reliability method. New York: John Wiley and Sons.

[14] Casciati, F, Colombi, P, Faravelli, L. (1993). Lifetime prediction of fatigue sensitive structural elements, *Structural Safety*, 12, 105-111.

[15] Casciati, F, Colombi, P, Faravelli, L. (1992). Fatigue crack size probability distribution via a filter technique. *Fatigue Fract. Eng. Mater Struct*. 15, 463-475.

[16] Yu, L. Purnendu, K.D, Zheng, Y. (2009). A response surface approach to fatigue reliability of ship structures. *Ships Offshore Struct*. 4, 253-259.

[17] Du, X. (2012). Reliability-based design optimization with dependent interval variables. *International Journal for Numerical Methods in Engineering*, 91, 218-228.

[18] Yang, H; Wang, A. (2012). Fatigue-reliability based design optimization of bending stiffener, *Journal of Ship Research*, 56,120-128.

[19] Chen, Z; Qiu, H; Gao, L; Su, L; Li, P. (2013). An adaptive decoupling approach for reliability-based design optimization. *Computers & Structures*, 117, 58-66.

[20] Dubourg, V; Sudret, B; Bourinet, J.M. (2011). Reliability-based design optimization using kriging surrogates and subset simulation, *Structural and Multidisciplinary Optimization*, 44, 673-690.

[21] Li, H.S. (2013). Reliability-based design optimization via high order response surface method. *Journal of Mechanical Science and Technology*, 27, 1021-1029.

[22] Noh, Y.J; Kang, Y.J; Youn, S.J; Cho, J.R; Lim, O.K. (2013). Reliability-based design optimization of volume fraction distribution in functionally graded composites. *Computational Materials Science*, 69, 435-442.

[23] Noh, Y; Choi, K.K; Lee, I; Gorsich, D; Lamb, D. (2011). Reliability-based design optimization with confidence level under input model uncertainty due to limited test data. *Structural and Multidisciplinary Optimization*, 43, 443-458.

[24] Song, C.Y; Lee, J. (2011). Reliability-based design optimization of knuckle component using conservative method of moving least squares meta-models, *Probabilistic Engineering Mechanics*, 26, 364-379.

[25] Leonel, E.D; Chateauneuf, A; Venturini, W.S; Bressolette P. (2010). Coupled reliability and boundary element model for probabilistic fatigue life assessment in mixed mode crack propagation. *International Journal of Fatigue*. 32, 1823-1834.

[26] Mi, Y. (1996). Three-dimensional analysis of crack growth. Topics in Engineering. V.28, Comp. Mech. Publications.

[27] Paris, P.C; Erdogan, F. (1963). A critical analysis of crack propagation laws. *Journal of Basic Engineering*, 85, 528-534.

[28] Pimenta, P.M.; Ricardo, L.C.H; Camarão, A.F. (2001). Modern trends on fatigue, SAE Brasil International Conference on Fatigue, São Paulo, Brazil.

[29] Portela, A; Aliabadi, M.H; Rooke, D.P. (1992). Dual Boundary element method: efficient implementation for crack problems. *International Journal of Numerical Methods in Engineering*. 33, 1269-1287.

[30] Hong, H. K; Chen, J.T. (1988). Derivations of Integral Equations of Elasticity, *Journal of Engineering Mechanics*, 114, 6:1028-1044.

[31] Hasofer, A.M; Lind, N.C. (1974). Exact and invariant second moment code format, *Journal of the Engineering Mechanics Division ASCE*. 100 (EM1), 111-121.

[32] Rackwitz, R; Fiessler, B. (1978). Structural reliability under combined random load sequences, *Computers & Structures*, 9, 489-494.

[33] Vanderplaats, G, N. (2001). Numerical Optimization Techniques for Engineering Design, Colorado Srings: Vanderplaats Research & Development.

[34] Onoufriou,T; Frangopol,D,M. (2002). Reliability-based inspection optimization of complex structures: a brief retrospective. *Computers & Structures*, 80, 1133-1144.

In: Recent Trends in Fatigue Design
Editor: Ricardo Branco

ISBN: 978-1-61668-410-5

Chapter 2

USE OF MODERN NUMERICAL METHODS FOR FATIGUE LIFE PREDICTIONS

Aleksandar Grbović and Boško Rašuo*
University of Belgrade, Faculty of Mechanical Engineering, Aeronautical Department, Serbia

ABSTRACT

In this chapter, FEM is used for fatigue crack growth predictions on 2D model of the spar, whereas XFEM is used for fatigue crack growth predictions and fatigue life estimation of corresponding 3D model. The values of stress intensity factors (SIFs) are extracted from FRANC2D software and MorfeoCrack for Abaqus add-on. The results obtained corresponded well with the experimental values, which proves that numerical methods can be successfully used in the determination of proper inspection intervals for the purpose of preventing the catastrophic failure of structural components under variable amplitude loading.

Keywords: Failure, Fatigue, Crack, FEM, XFEM, Aircraft

---

* Corresponding author: Boško Rašuo. Address: Kraljice Marije 16, 11120 Belgrade 35, Serbia. E-mail: brasuo@mas.bg.ac.rs.

INTRODUCTION

In recent decades the assessment of the service durability of mechanical components and assemblies has become an important segment of design, mostly because of the growing needs for light-weight structures which will be safe and reliable, and at the same time, not too expensive. It is especially true for the main structural elements used in the aircraft industry such as wing spars and fuselage bulkheads and fittings, whose sudden failure could lead to catastrophic consequences. In order to meet strict safety requirements, as well as to check structural components before usage, a number of expensive and long experiments are carried out. Taking into account ever-present manufacturers′ tendencies to shorten time-to-market periods, the use of finite element method (FEM) and extended finite element method (XFEM) for the estimation of fatigue life has been proven as a viable alternative to the experimental methods. The purpose of this chapter is to show how FEM and XFEM could be used for obtaining not only an accurate estimation of the fatigue life of the assembly (such as the spar of a light aircraft), but also a reliable prediction of a number of load cycles which would cause a crack in the spar, and the length of such. On the basis of these results, it is possible to determine the proper inspection intervals which could prevent a catastrophic failure of the structure under variable amplitude loading.

Standard specimens tested under constant amplitude loading have given data on how materials behave under different fatigue conditions [1].

However, during service the structural components of aircraft are subjected mostly to the variable amplitude loading. This means that available fatigue data from tests under constant amplitudes cannot be applied without appropriate modifications. Furthermore, the geometry of real components is very often significantly different from the geometry of specimens, which could highly influence the accuracy of fatigue life predictions.

In such cases the experimental verification of fatigue life of components (or assemblies) under variable amplitudes of loading must be carried out.

But even the smallest change in the loading geometry of materials of components necessarily leads to new experiments, making the design process more expensive.

In a case when a numerical model of the structural component or assembly is developed all required changes are relatively easy to implement. Then new estimations of fatigue life can be obtained quickly and at low costs. However, the question that must be answered is: how can we be sure that the numerical model is valid i.e. that results are reliable and acceptably precise?

It is obvious that an initial numerical model must be experimentally verified, and after that the values of fatigue life, obtained by calculations for variable amplitude loading, may be considered as good enough.

Of course determination of absolutely accurate values is impossible because of the uncertain nature of fatigue failures, i.e. identical components, subjected to the same loading spectrum during the experiment very often experiencing large variations in fatigue life [2].

However, numerical methods at least help us to define criteria for acceptance or rejection of the design and also allow Concept A to Concept B comparisons to be made without the need for complete accuracy.

The most important roles of the aircraft structures during their lives are to transmit and carry loads and provide required lift force (mostly generated by airfoils moving through the air). This is achieved by using thin-walled structures in which the interior surfaces are reinforced by longitudinal and transverse strengthening members. A predominant method used for joining these elements is riveting.

The wing represents one of the most important assembly zones in aircraft structures and special attention must be paid to determine wing fatigue life. The most important element of the wing is the I-beam (spar) which provides stability under compressive loads. It is extended along the length of the wing, in a direction perpendicular to the fuselage.

Figure 1. Light aircraft UTVA 75.

Normally, the wing has 2-3 spars, but light aircraft – such as UTVA 75 shown in Figure 1 – mostly have wings with one spar [3].

The spar carries almost all of the bending and shears loads, of which the lift force is the most dominant. Generated lift force bends the whole wing upright; as a consequence, the upper elements of the spar are under compression, while the lower are under traction.

The subject of crack growth analysis in this case is the I-beam (shown in Figure 2) of the UTVA 75 aircraft. A critical zone in which, under service loading, the crack is most likely to appear, is located in the wing root on the lower side and covers the spar caps and the wing skin. In this case, spar caps and spar web are connected with two rows of 3.2 mm diameter rivets.

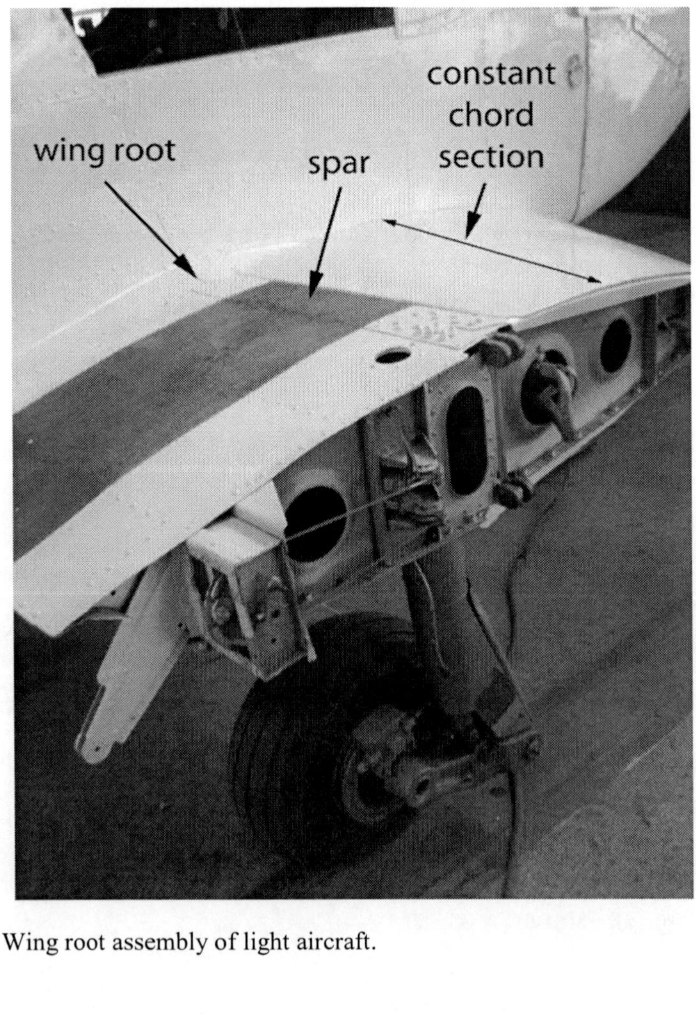

Figure 2. Wing root assembly of light aircraft.

Spar assembly used in fatigue testing, as well as washers and supporters used to fix spars and apply the load during the experiment, are shown in Figure 3. The spar without adjoining skin was tested, so the reinforcing effect of the wing skin was not taken into consideration.

The spar parts are mainly made of hardened aluminum alloys. The main advantages of aluminum alloys are lightness, high specific strength and good corrosion resistance. The alloys that provide major mechanical properties are Al-Zn-Mg-Cu (the 7xxx family) and Al-Cu-Mg (the 2xxx family). The two mostly used alloys in aerospace applications are 7075-T6 and 2024-T3 [4]. The spar caps in Figure 3 were made from 2024-T3 aluminum plates (thickness 1.6 mm), while the spar web was made from the same material, but of different thickness (1 mm). Chemical composition of 2024-T3 aluminum is shown in Table 1 [5]. For the purpose of fatigue testing, ten identical spars have been made [6]. The length of the spars was 600 mm which equals the length of the constant chord section shown in Figure 2. In order to measure stress and strain in the zone where crack initiation was expected, several strain gauges were used. (Strain gauge location is shown in Figure 4).

One of the fundamental issues in the reliability analysis of fatigue behavior of real components under service loading is the evaluation of load or stress spectra.

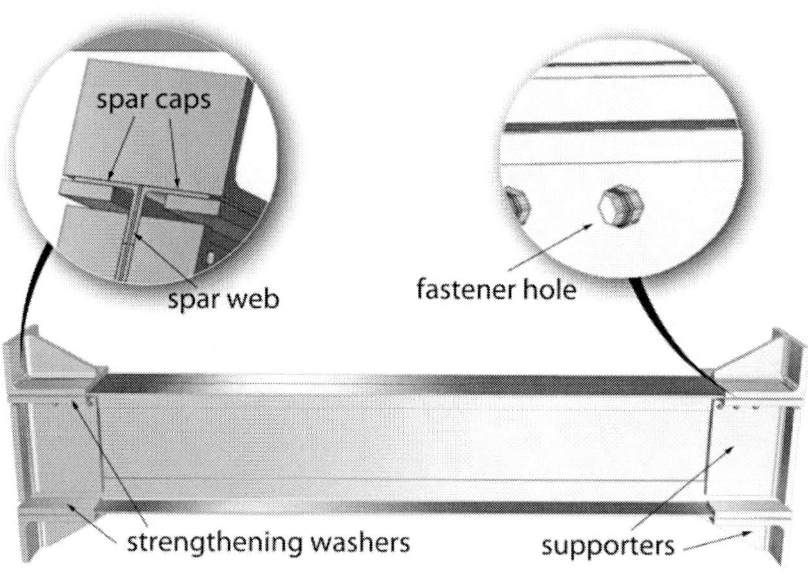

Figure 3. The spar and supporting elements used in experiment.

It is a known fact that the nature of the service load is stochastic and that data from original operating conditions are often not available [7-9]. The aircraft service load history is a mixture of random loads and deterministic loads (such as loads in maneuvers). In the past, only deterministic loads were used in experiments, with random loads being represented by one or two amplitudes. However, the modern principle of flight simulation fatigue tests (for problems relating to the tension of aircraft wing structures) is completely different and is shown in Figure 5 [10].

The curve in Figure 5 represents the superposition of deterministic loads and two types of random loads: turbulence (gust loads) and taxiing loads. At cruising altitude, gust loads rarely occur. At lower altitudes turbulence is a matter of weather conditions and gusts vary from flight to flight.

Table 1. Chemical composition of aluminum alloy 2024 T3

Component	W_t (%)	Component	W_t (%)
Al	90.7-94.7	Cr	Max.0.1
Cu	3.8-4.9	Fe	Max. 0.5
Mg	1.2-1.8	Mn	0.3-0.9
Si	Max.0.5	Ti	Max. 0.15
Zn	Max.0.25	Other, each	Max.0.05
		Other, total	Max. 0.15

Figure 4. Location of strain gauges used in experiment.

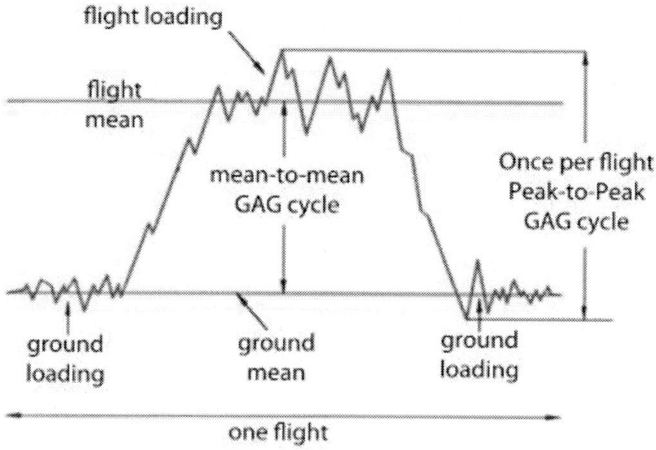

Figure 5. Ground-Air-Ground (GAG) load history used in fatigue tests.

Therefore it is normal to simulate some eight to ten different weather conditions in a flight simulation fatigue test [11]. During landing and takeoff taxiing loads occur as a result of runway roughness and if they are accepted as non-damaging they may be omitted.

In recent decades, several standardized load histories have been developed for the purpose of simplification and standardization of fatigue testing procedures. Different load histories have been defined for different types of aircraft (for example, TWIST and mini-TWIST are used in transport aircraft testing, FALSTAF and mini-FALSTAF for fighter aircraft, etc.).

Available data may also be used for aircraft types with no defined standardized histories because the sequences of loads in standardized histories are completely defined by numerical format and the load scale is selectable.

The main problem with the use of standardized spectra in fatigue life estimations is the omission of numerous small cycles because if these cycles were included tests would take an unacceptably long time.

In a fatigue test the load variations must be applied at a higher loading rate than in service (usually at a rate of ten flights per minute) [12].

Hence, in order to get reliable values of fatigue life, it is necessary to test the component or the whole assembly under variable amplitude loading by using either the available service load history (from the aircraft of the same type) or by using a standardized spectrum.

The increased complexity of the loads history that must be simulated requires the development of expensive computer-controlled systems.

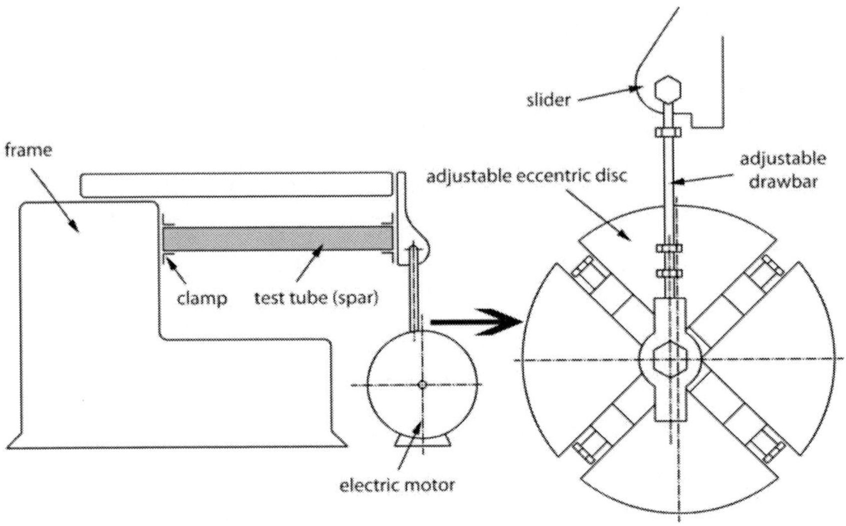

Figure 6. Fatigue system used in testing of light aircraft spar.

Figure 7. Main components of fatigue testing system.

A much simpler and cheaper system (shown in Figures 6 and 7) was developed for the purpose of fatigue testing of light aircraft spars. This system can simulate different and complex load histories by changing the position of an adjustable drawbar and the configuration of an adjustable eccentric disc.

During the experiment, the force on the drawbar was measured using a force transducer of type U3 50 kN, manufactured by Hettinger Baldwin Messtechnik, Germany (Pos. 1 in Figure 7). Assuming the drawbar to be rigid under a predicted external force, deflection of one end of the spar was measured relatively between the spar′s end and center point of the drawbar, via a potentiometer displacement transducer PM2S 150 mm, manufactured by ELAP Spa, Italy (Pos. 2 in Figure 7). Stresses were measured in the clamp area, at four points, using aluminum strain gauges of the type 1-LY13-10/120, produced by Hettinger Baldwin Messtechnik (Pos. 3 in Figure 7). Each strain gauge was connected to one dummy strain gauge (Pos. 4 in Figure 7) into one half of a Wheatstone bridge circuit, and then with the acquisition system. Data acquisition and analysis were performed using a SPIDER8 measurement acquisition system (Figure 8) and software package Catman Express, produced by Hettinger Baldwin Messtechnik.

Firstly, the spars were tested under constant amplitude loading. On the basis of obtained values, the numerical model was checked.

Figure 8. Acquisition system Spider8.

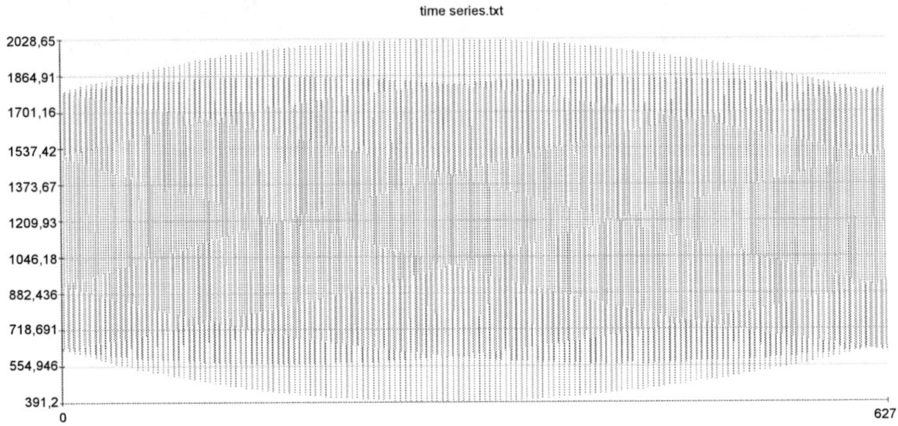

Figure 9. Narrow band random loading used in experiment and FEA.

An initial finite element model of the light aircraft spar (Figure 10), including supports and washers, was developed by the Ansys Workbench software [13].

The final model used in the calculations had 2,410,012 nodes and 1,224,900 finite elements and was obtained by means of an iteration process which consisted of comparison of numerical values obtained for current mesh against the values obtained in tests under constant amplitude loadings (four spars being used for that purpose).

In each step of the iteration process the mesh was improved in order to get numerical solutions closer to the experimental values. Fatigue life was calculated using the stress-life (S-N) method, because the number of cycles to visible damage in the experiments was over $3x10^6$ for all four spars. When the fatigue life of the spar, obtained by finite element analysis, was finally very close to the mean value of the experimental results the numerical model was accepted as satisfactory. Tests of the spar under variable amplitude loading started immediately after that; narrow band random loading, shown in Figure 9, was one of the load histories used in the tests.

Narrow band random loading was produced on the apparatus shown in Figure 6 in such a way as to cause damage on upper caps; the reason is that it´s much easier to see cracks on the upper surface of the spar and then to observe its growth. Load history used in experiment, (minimum value +391.2 N, maximum value +2028.0N, frequency 12.5 Hz), produced strong tensile stress in upper caps, so visible cracks appeared on the left cap below the support, after only 8,542 cycles (Figure 11).

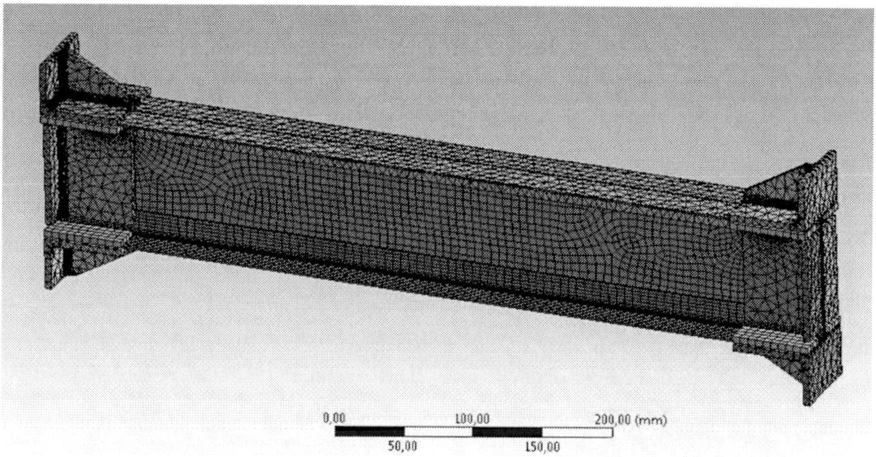

Figure 10. Initial finite element mesh.

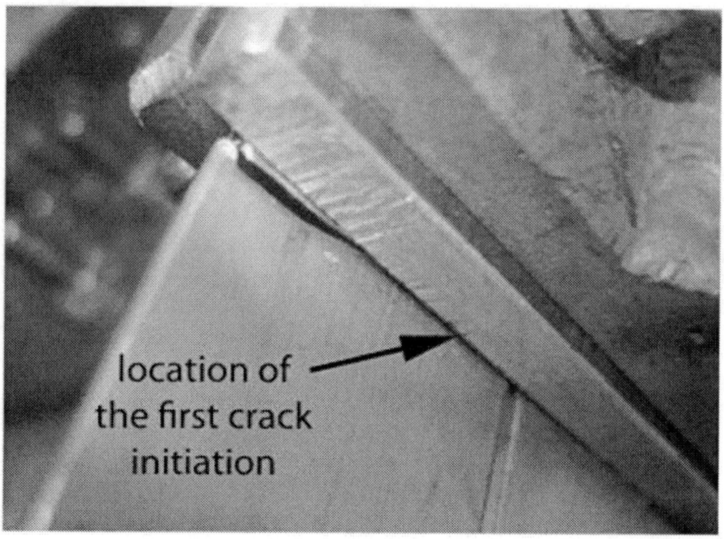

Figure 11. Location of the first crack initiation in the experiment.

The crack began to spread rapidly towards the spar web, then changed its direction and continued along the cap at an angle of 90° with respect to the original direction (Figure 12). Simultaneously, it moved towards the left edge of the cap. In order to carefully study the crack propagation, testing was paused at intervals, some inspections were carried out and the current length of crack was recorded.

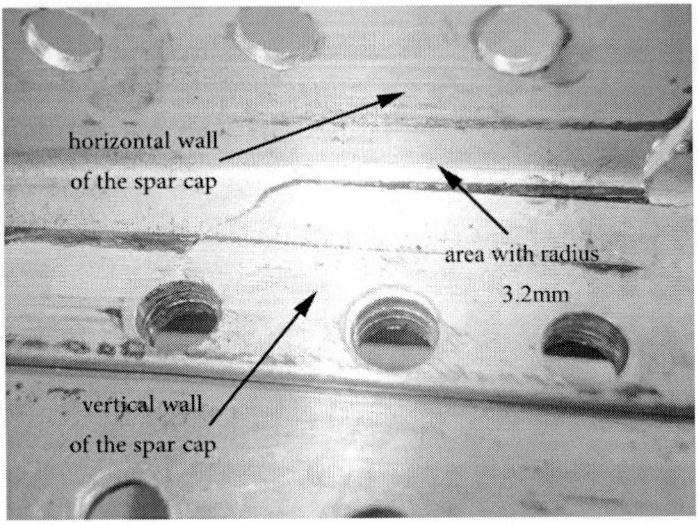

Figure 12. Path of the 1st crack on vertical wall of the spar cap.

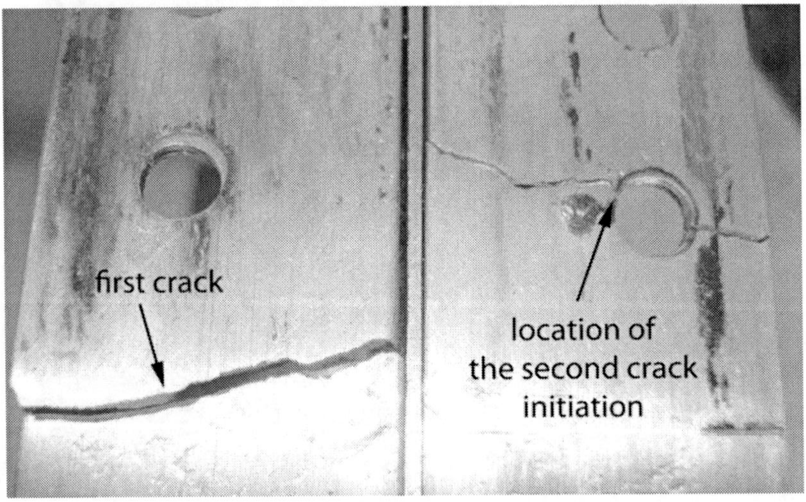

Figure 13. Locations and paths of cracks on horizontal walls of the spar caps.

After 39,450 cycles, another crack was spotted on the right spar cap, later determined to have occurred on the fastener hole (Figure 13).

The crack was spotted when it moved to the area between the horizontal and vertical walls of the cap, and because it occurred below the upper support its initiation wasn't registered visually.

Cracks then continued to develop below the strengthening washers but were not visible, so after 58,520 cycles the test was stopped and the spar was examined in order to determine what was going on below the strengthening washers. It was noticed that both cracks continued to the side fastener hole, i.e. to the last holes on both upper caps (Figure 14 and 15).

It was also found that the crack didn't appear on the spar web and – aside from the cracks on the caps – at the moment when the experiment was stopped there were no signs of any other damage on the spar assembly (Figure 16).

After the experiment a FE analysis was performed in order to determine if the numerical model in Ansys, under the loading history taken from Spider8, could predict a value of the fatigue life close to actual life. Taking into account that the first crack appeared after only 8,542 cycles the ε-N method (strain life method) was used in the calculation, with the spectrum (Figure 9) being loaded from Spider8 into Ansys. As both the variable amplitude loads and displacements were measured during the experiment, calculations in Ansys were performed not only with the variable force acting at the end of the spar, but also with applied displacement at one end of the spar.

Figure 14. First crack growth path (on the vertical wall of the left spar cap).

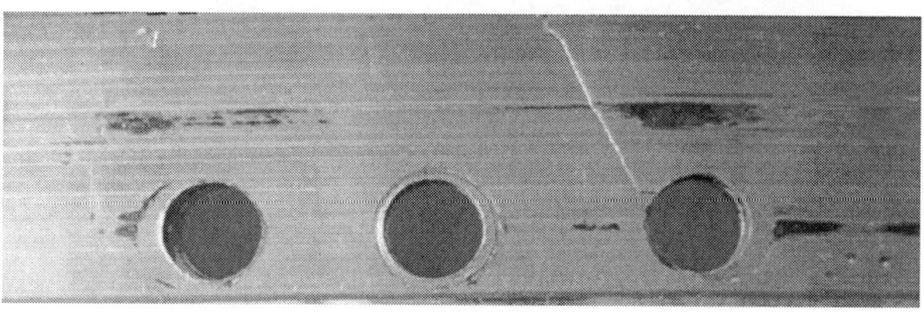

Figure 15. Second crack growth path (on the vertical wall of the right spar cap).

Fatigue life values obtained in both cases were slightly different: 50.6 blocks in the case of applied force (Figure 17) and 52.6 blocks for applied displacement. Considering that one block consists of 157 cycles, it is easy to calculate that FEA predicts that the crack will appear on the spar after approximately 7,944 cycles. In Figure 17, it is clear that Ansys predicts initiation of the crack near the location where it was detected in the experiment, confirming that the numerical model was well defined. The strain life method in Ansys provides two mean stress correction methods (Morrow mean stress and SWT [11]) and after initial calculations both methods were applied. (They didn't show any significant difference in results).

After having obtained approximately the same number of the cycles to initial damage, a new numerical model was defined, for the purpose of performing crack growth simulation in both upper spar caps. Considering that both cracks moved firstly along the horizontal walls of the caps and then along the vertical walls (connected to spar web by rivets), it was concluded that the crack growth should be simulated from the moment they started to develop along the vertical walls towards the side fastener holes. It was estimated that this was the critical phase of growth for both cracks. Simulation of crack growth on horizontal wall was not challenging because of only one layer of material under loading.

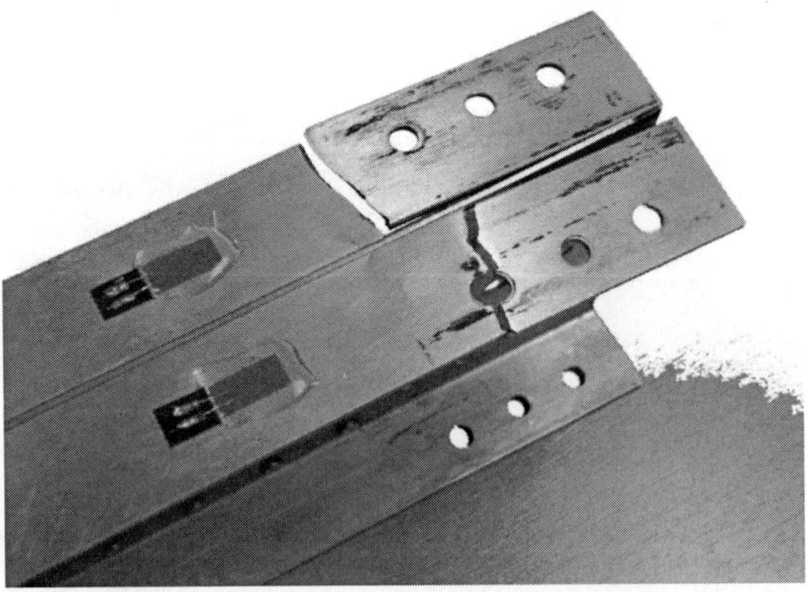

Figure 16. Spar after experiment with narrow band random loading.

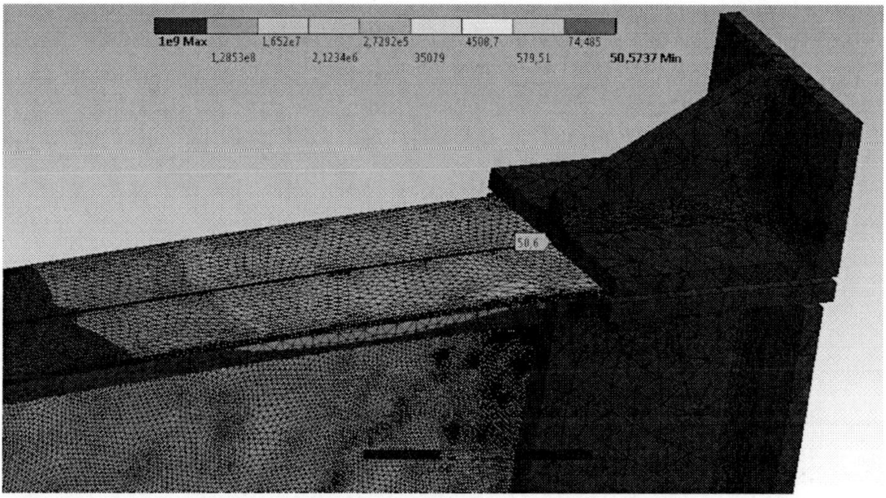

Figure 17. Values of fatigue life (number of blocks of spectrum) obtained in Ansys.

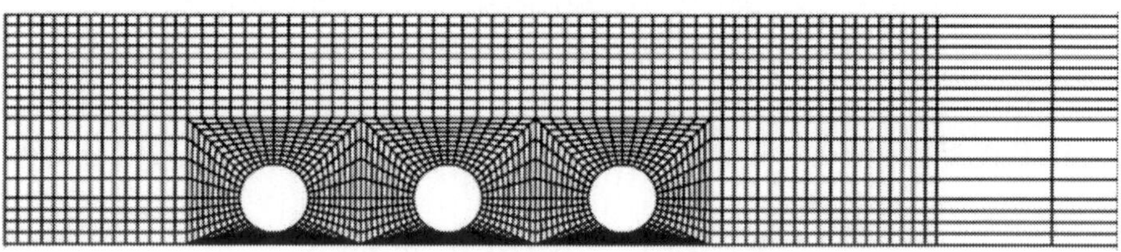

Figure 18. Finite element mesh around fastener holes on upper spar caps (1st and 3rd layer).

On the other hand, analysis of the crack growth on vertical walls includes a definition of three layers of materials with different thicknesses (two caps and web) and simulation of riveted joints. Since the thickness of caps is 1.6 mm and the thickness of the web is 1 mm, this can be considered as the through crack case and numerical modeling of crack propagation in a 2D aluminum plate should provide good results.

Numerical modeling of the through crack was performed in the software for 2D crack analysis FRANC2D/L [14].

FRANC2D/L is an interactive program for crack growth simulations in multi-layered structures and provides representation of lap joints or bonded repairs.

Each layer must be represented by a separate mesh that, in the zone of overlapping, must be identical to the mesh of other layers and which is connected to them with rivets or adhesive elements.

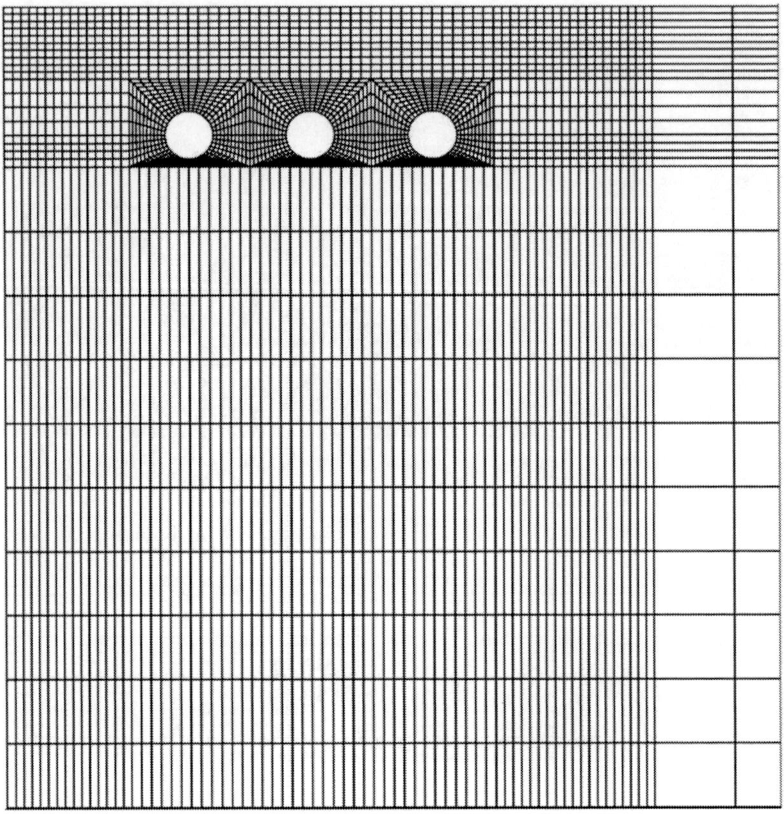

Figure 19. Fragment of finite element mesh of the spar web (2nd layer).

Aside from that, all layers must be flat and both two-dimensional and three-dimensional calculations can be performed. Figure 18 shows a part of the generated mesh of the first and third layers (thickness 1.6 mm) around the fastener holes on the vertical walls of the upper spar caps.

The second layer (shown in the Figure 19) represents the spar web (thickness 1 mm) and it is placed between the first and third layer. If Figures 18 and 19 are compared it is noticeable that the mesh, in all three layers, is identical in areas where caps are connected to the web; a condition which has to be met in order to connect all layers with rivet elements in FRANC2D/L. The mesh of all three layers was made using the CASCA program [14] and the layers were connected by rivet elements.

In the FRANC2D/L program the rivets are considered as elastic shear springs where the rivet stiffness is the constant of material (rivet stiffness can be determined by approximate formula $K_{rivet}=12EI/7L^3$ [15]).

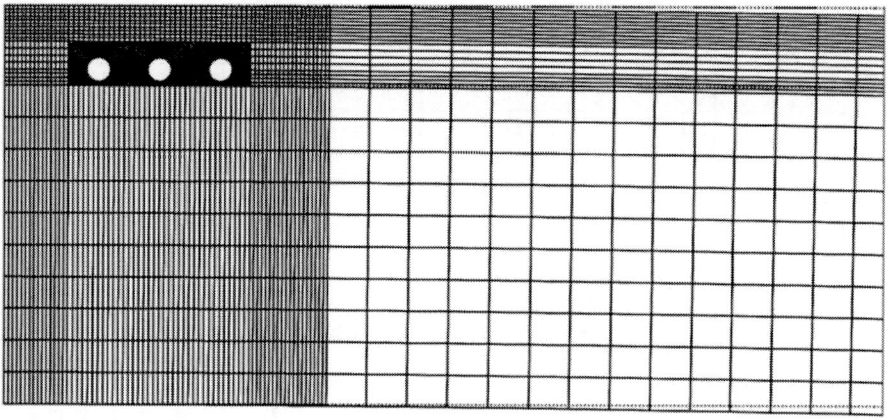

Figure 20. Deformed meshes of all three layers.

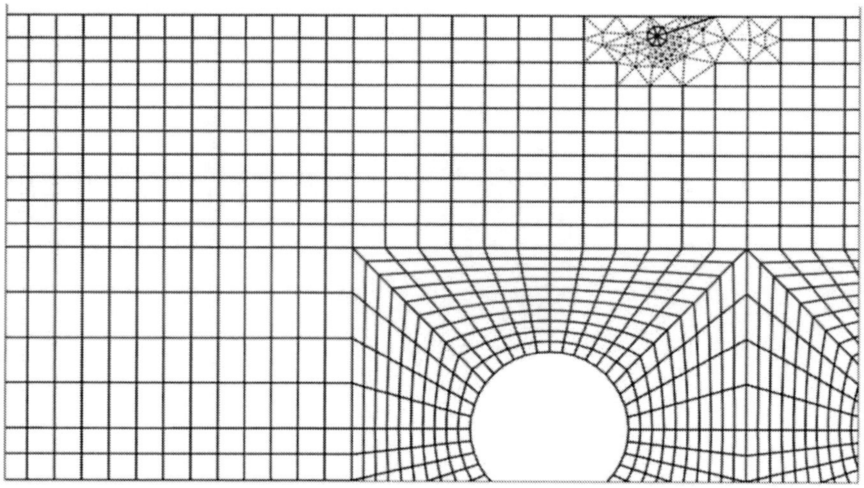

Figure 21. Initial position of the first crack.

Rivets may only be introduced at nodes and when they connect two layers, they must be defined at a node in the upper layer. This eliminates the need for interpolation of displacements at points where rivets were defined, in order to calculate the forces on the rivets.

On the meshes in Figures 18 and 19, rivet elements were defined in the nodes that were very close to the positions of real rivets on the spar used in the experiment. The lower spar caps were not modeled in order to reduce the number of finite elements used in the calculations.

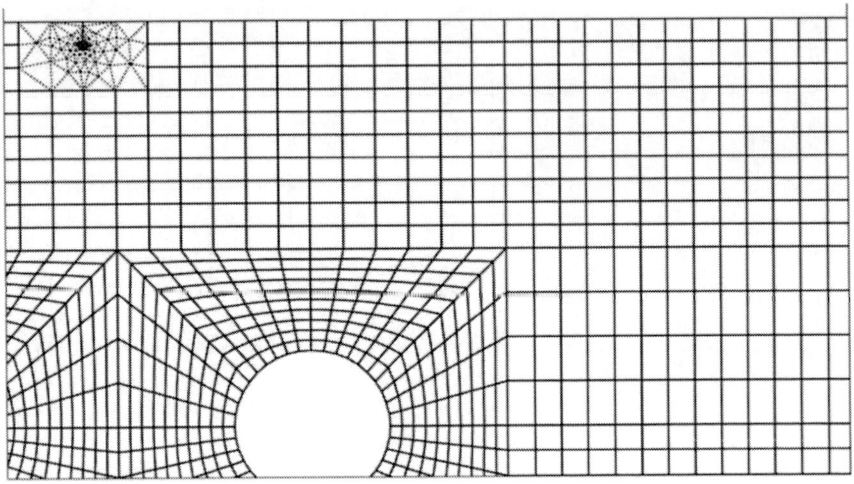

Figure 22. Initial position of the second crack.

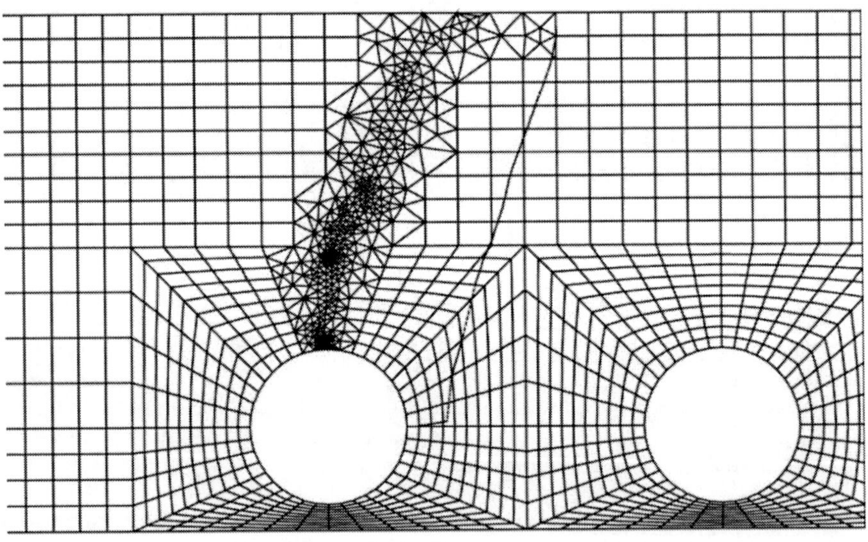

Figure 23. First layer mesh after crack growth (dashed line represents second crack path).

After the layers had been connected displacements of nodes on the left edges of all three layers were fixed in both directions as well as displacements of nodes at the upper and lower edges in the clamp area. Then a 3 mm displacement in the vertical direction (maximum displacement in experiment) was applied on all nodes of the right edge of the second layer (spar web).

Deformed meshes of all three layers (shown in Figure 20) were the proof that rivet elements successfully transferred the loading from layer 2 to layers 1 and 3. Values of stresses obtained in FRANC2D/L were close to the values obtained in Ansys.

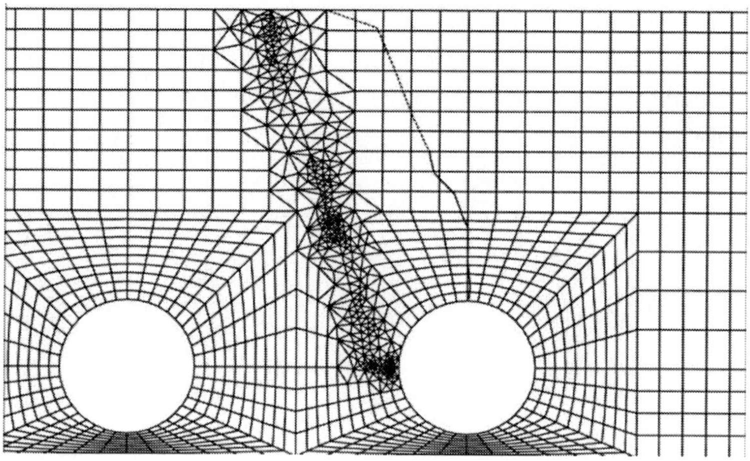

Figure 24. Second layer mesh after crack growth (dashed line represents first crack path).

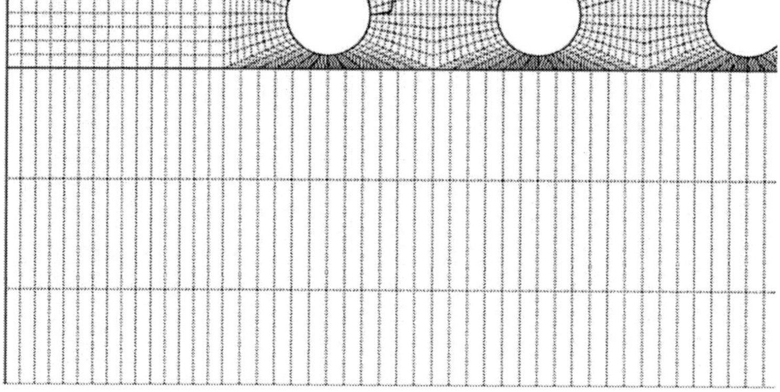

Figure 25. Cracks′ paths seen from the second layer.

Since in FRANC2D/L the user can choose between two types of cracks, internal and edge, it was decided to initiate both cracks on the upper edges of the first and third layers and to develop them towards the fastener holes. To determine the exact positions for the cracks' beginning, Figure 14 and 15 were used. The initial position of the crack on the first layer is shown in Figure 21 whereas the position of the second crack is shown in Figure 22. Stress contours didn't indicate the highest values of stress on these spots, but during the experiment cracks moved from horizontal caps´ walls to vertical walls at these positions.

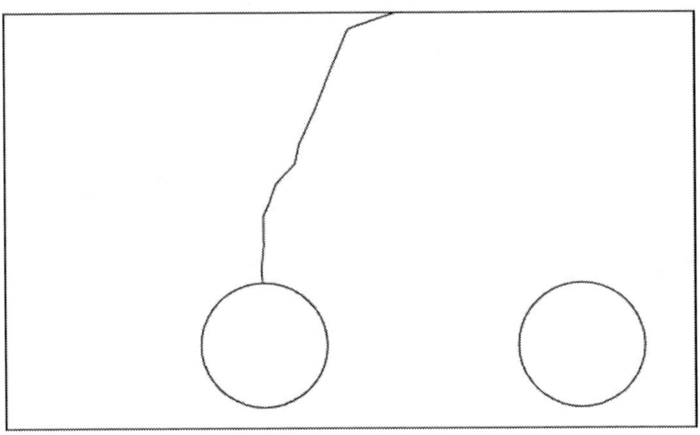

Figure 26. First crack path without finite element mesh.

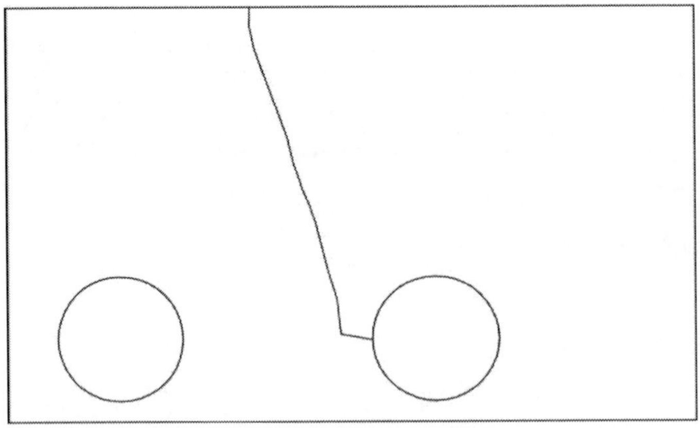

Figure 27. Second crack path without finite element mesh.

The cracks were developed using a standard method that predicts direction of propagation according to maximum hoop stress in the vicinity of the crack tip [16]. Each time cracks were produced the mesh was modified in order to reflect the current crack configuration and values of stress intensity factors were calculated using the displacement correlation method or modified crack closure techniques [16]. The cracks were propagated simultaneously and after nine steps of propagation they came to the edges of the fastener holes (Figure 23 and 24).

The stress intensity factor history for both cracks was saved as a text file. In Figure 25 the cracks′ paths are shown as seen from the second layer, whereas in Figure 26 and Figure 27 the cracks are shown without the finite element mesh. Comparing Figures 26 and Figure 27 to Figures 14 and 15 it can be seen that paths obtained in finite element simulation were similar to those observed in the experiment with variable amplitude loading, shown in Figure 9.

The stress intensity factors (SIFs), saved in a text file, were used as input to FRANC2D/L Crack Growth Program (F2D/LCG) that permits the calculation of the crack growth curve. Other inputs for F2D/LCG are Time History File (loading spectrum) and Material File (mechanical and fatigue properties of 2024-T3 aluminum alloy shown in Table 2 [5]).

The spectrum is counted by F2DL/CG with a Range-pair algorithm [17], and the crack growth curve is calculated by integrating a modified Forman-Mettu NASGRO equation for fatigue crack growth rate (da/dN) expressed as a function of crack-tip stress intensity factor range ΔK [18]:

$$\frac{da}{dN} = C\left[\left(\frac{1-f}{1-R}\right)\Delta K\right]^n \frac{\left(1-\dfrac{\Delta K_{th}}{\Delta K}\right)^p}{\left(1-\dfrac{\Delta K_{max}}{\Delta K_{crit}}\right)^q} \tag{1}$$

In this equation C, n, p and q are the material parameters, f denotes the crack opening function, whereas R-ratio of the load is defined as $R=R_{min}/R_{max}$ (or $R=K_{min}/K_{max}$). F2D/LCG has set of files providing NASGRO equation coefficients for aluminum alloys, and after selecting 2024-T3 aluminum (thickness 1.6 mm) da/dN curve is shown (Figure 28).

Table 2. Mechanical and fatigue properties of aluminum alloy 2024-T3

Title	Symbol	Value
Yield Stress (MPa)	Q_y	365.4
Ultimate tensile Strength (MPa)	Q_u	455.1
Plane Strain Fracture Toughness (MPa mm$^{1/2}$)	K_{IC}	1042
Part Through Fracture Toughness (MPa mm$^{1/2}$)	K_{IE}	1459
Paris Exponent	N	3.20
Forman Coefficient	C	2.382×10^{-12}
NASGRO Exponent	P	0.25
NASGRO Exponent	Q	1
Modulus of Elasticity (GPa)	E	71.75
Fatigue Strength coefficient (MPa)	f	130

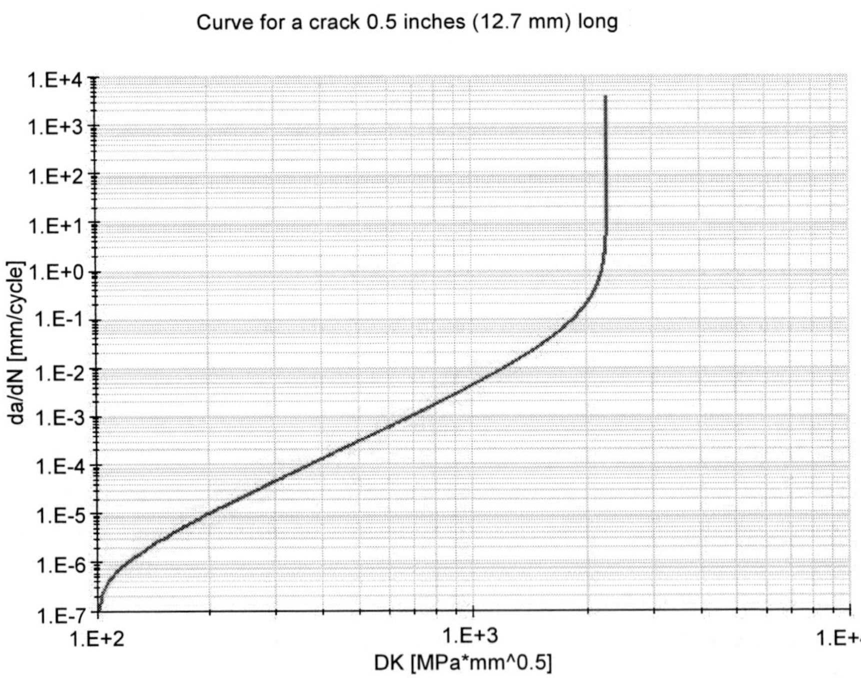

Figure 28. Aluminum 2024-T3 d*a*/*dN* *vs.* *ΔK* curve (displayed curve is for a specimen with a crack length of 12.7 mm).

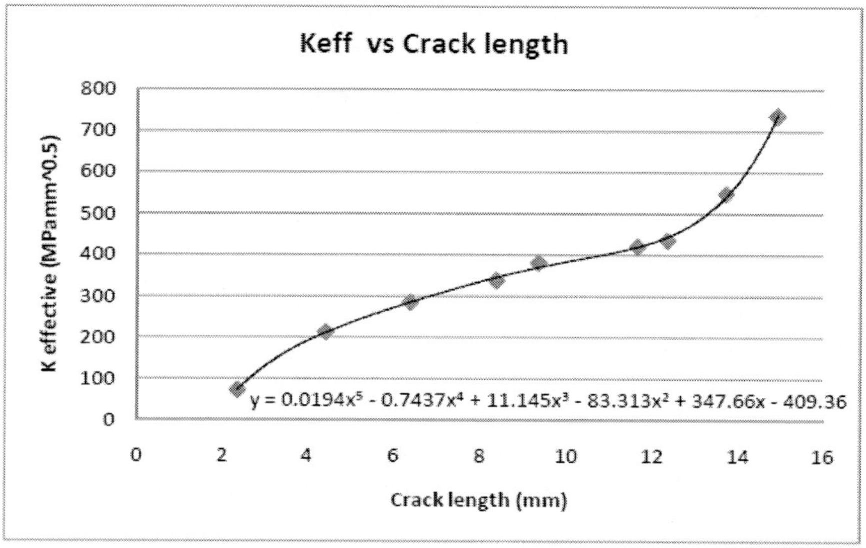

Figure 29. K_{eff} vs. Crack length for first crack (polynomial order is 5).

The thickness of material is one of the independent parameters entering in the NASGRO equation and it affects the critical SIF (ΔK_{crit}) according to the equation (2):

$$\frac{\Delta K_{crit}}{K_{IC}} = 1 + B_K \cdot e^{-\left(\frac{A_k t}{t_0}\right)} \tag{2}$$

where K_{IC}, A_K and B_K are material parameters, while t_o is given by the following equation (in plane strain conditions):

$$t_0 = 2.5 \cdot \left(\frac{K_{IC}}{\sigma_{ys}}\right)^2 \tag{3}$$

By default, the Mode I SIF (K_I) is used for integration of Equation 1, so ΔK is obtained according to the formula $\Delta K = K_{I\ max} - K_{I\ min}$. However, integration on the basis of K_I values gave the fatigue life N=124,492 cycles for the first crack. This value is almost three times as big as the value obtained in the experiment. For second crack calculated fatigue life was N=14,001 cycles.

Analysis of the SIF history file showed that the values of stress intensity factors Mode II (K_{II}) were not negligible for the first crack. On the basis of this finding it was determined to use K_{eff} in calculation, given by equation (4):

$$K_{eff} = \sqrt{K_I^2 + \beta \cdot K_{II}^2} \tag{4}$$

where ß is material property. Therefore the ΔK used in Equation 1 is $\Delta K = K_{eff\,max} - K_{eff\,min}$.

In order to perform a new analysis, values of K_{eff} were calculated for each step of crack propagation. After that, the polynomial fit was calculated. The equations of the lines of best fit obtained for both cracks, as well as graphs K_{eff} vs. Crack length, are shown in Figure 29 and Figure 30.

Instead of K_I, K_{eff} was used for integration of Equation 1 and results showed that the number of cycles required for first crack to grow up to a=14.92 cm was N=40,413 (i.e. 257.4 blocks of spectrum shown in Figure 9) and that the number of cycles required for the second crack to grow up to a=13.26 cm was N=13,392 (i.e. 85.3 blocks). Graphs Crack length vs. Life are shown in Figure 31. Calculated fatigue life values are similar to values obtained in the experiment which justifies the use of K_{eff} in calculation.

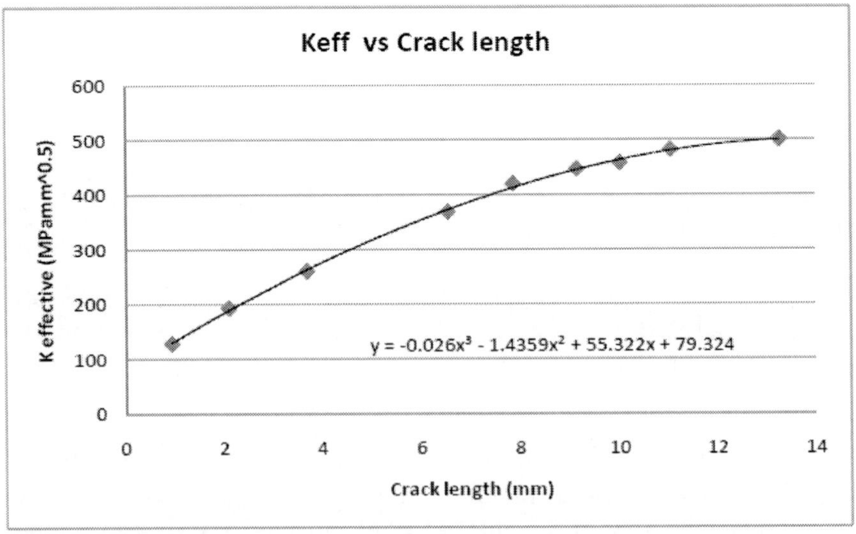

Figure 30. K_{eff} vs. Crack length for second crack (polynomial order is 3).

It is especially true for the first crack, whereas the difference in number of cycles obtained for K_I and K_{eff} in the case of the second crack is very small (14,001 versus 13,392 cycles). This indicates the negligible influence of K_{II} on the second crack growth rate.

Finally, the crack growth curve calculation was performed for standard spectra mini-TWIST and mini-FALSTAFF [19] (Figure 32 and Figure 33) in order to obtain the estimation of the number of cycles required for cracks to grow up to the dimensions obtained in the experiment.

The calculation in Ansys under the mini-TWIST spectrum showed a damage appearance after approximately 660,000 cycles and under mini-FALSTAFF spectrum after 290,000 cycles.

Figure 34 shows the fatigue life N=1,849,468 cycles (or approximately 103 flight hours) obtained for the first crack under the mini-TWIST spectrum loading, as well as the fatigue life N=1,072,496 cycles (or approximately 60 flight hours) for the same crack under the mini–FALSTAFF spectrum. In Figure 35 obtained values of fatigue life for the second crack can be seen: under mini-TWIST spectrum N=389,350 cycles (or 22 flight hours) and under mini-FALSTAFF spectrum N= 257,220 (or 14 flight hours).

Comparing the values obtained in the experiment against those obtained by numerical simulation, the similarity of results is obvious. In the experiment, the first crack appeared after 8,542 cycles and FEA predicted damage appearance under a variable amplitude load not below 7,944 cycles.

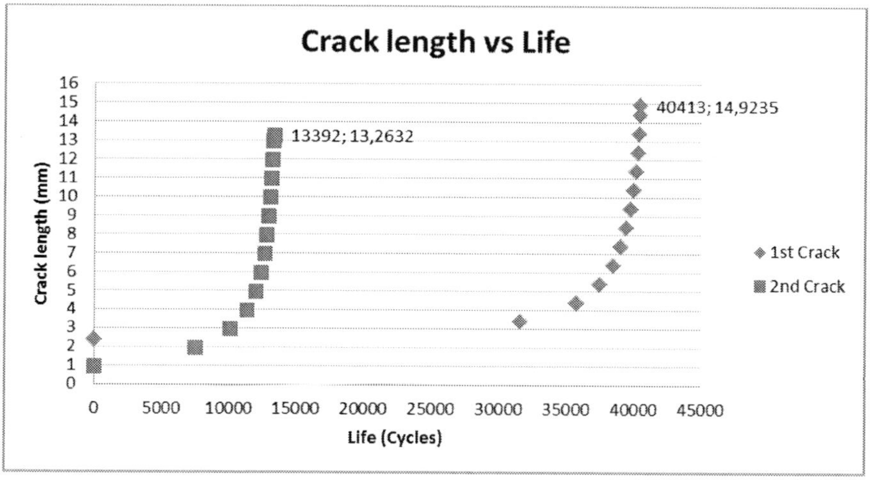

Figure 31. Graph Crack length vs. Life for both cracks.

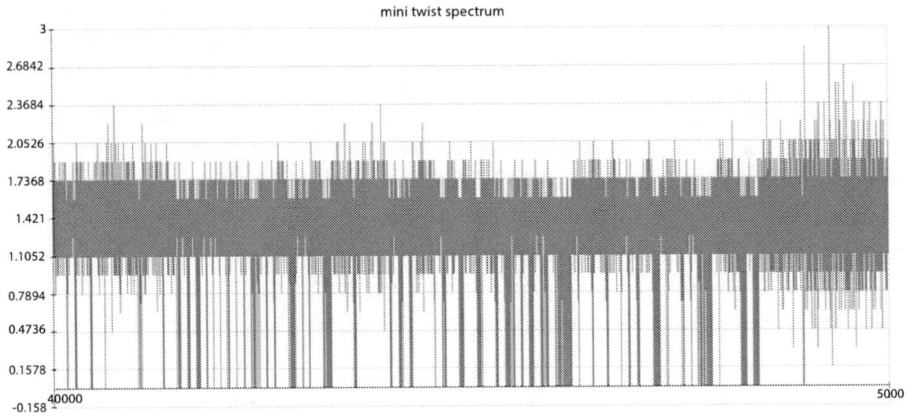

Figure 32. Mini-TWIST spectrum used in calculations.

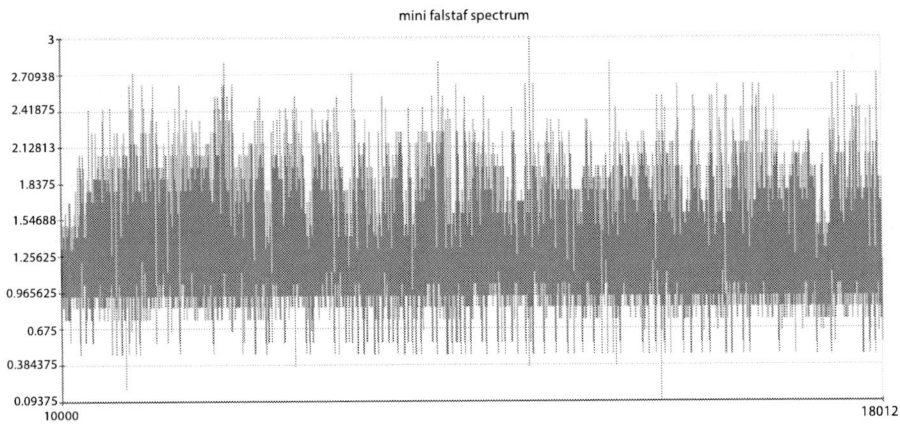

Figure 33. Mini-FALSTAFF spectrum used in calculations.

Adding this value to the number of cycles N=40,413 required to move the first crack on the left spar cap wall towards the last fastener hole, we get a total fatigue life of N=48,357. In the experiment the first crack reached the last fastener hole after 58,520 cycles.

The difference of approximately 10,000 cycles may be explained by the fact that before the crack moved from the horizontal to the vertical wall of the spar cap it had spread for a while through the area between the vertical and horizontal walls.

The second crack was seen later in the experiment (after 39,450 cycles), at the moment when it became visible in the area between the horizontal and vertical walls.

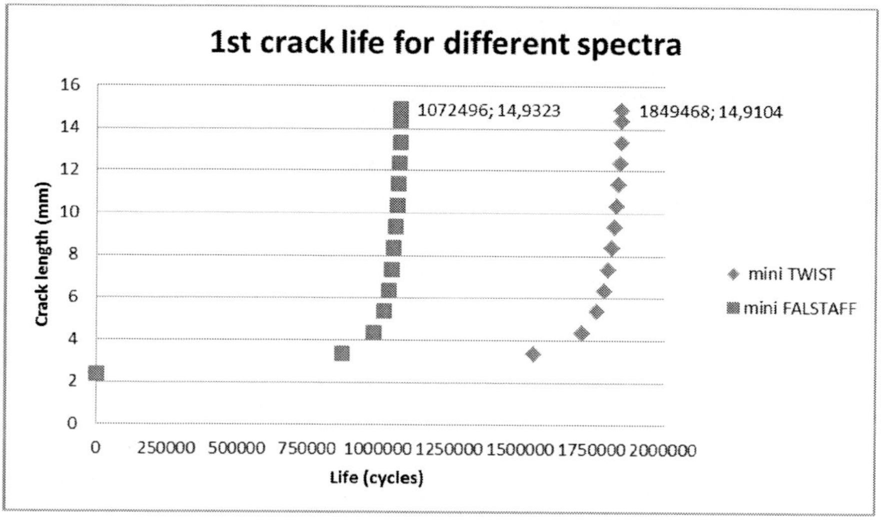

Figure 34. First crack length vs. Life under different spectra.

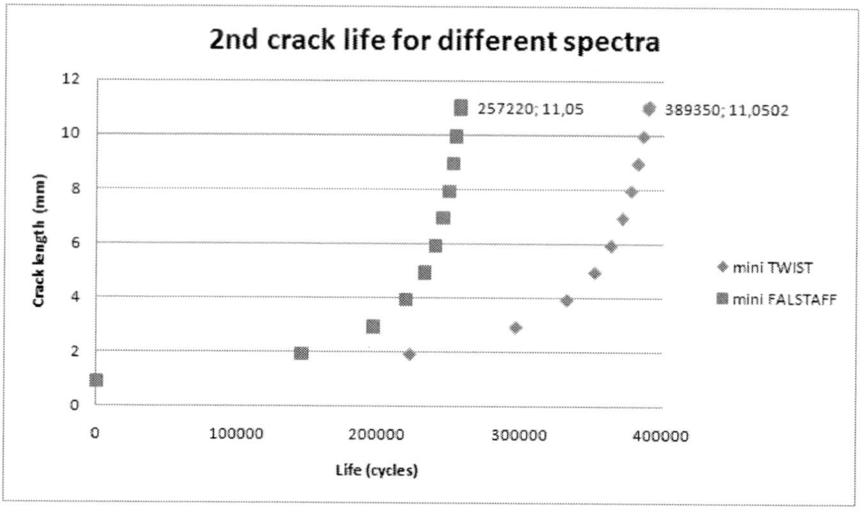

Figure 35. Second crack length vs. Life under different spectra.

It was found that this crack began from the fastener hole below the upper support and this may be explained by the stress concentration around a circular hole. Unfortunately, the exact number of cycles required to initiate the second crack wasn't recorded. Adding the calculated number of cycles (N=13,392) required to move the second crack on the right spar cap wall towards the fastener hole to the number of cycles to the visible crack in the experiment, we get total fatigue life N=52,842 for the second crack. As previously suggested, the difference of approximately 6,000 cycles can be explained by crack development in the area between the walls.

Values of fatigue life under mini-TWIST and mini-FALSTAFF loading spectra were relatively low, but it must be noted that these are standard spectra used in transport and fighter aircraft fatigue tests, not in light aircraft tests. However, the light aircraft service load history is in some ways similar to the transport aircraft load history. Values of fatigue life under the mini-TWIST spectrum obtained by numerical simulation are between several hundred thousand cycles (second crack) and several million cycles (first crack). In practice, when aircraft inspection intervals have to be determined, these values can be considered as reliable.

Finally, the entire process of the crack growth simulation by means of FEM has proved successful and reliable because calculated values matched the results of the experiment. Of course it must not be forgotten that fatigue has a stochastic nature and that the same spar under the same loading may show different fatigue behaviors, i.e. the crack may occur sooner or later, and on other locations too. However, numerical models may be used for new crack growth simulations and estimations of the residual life of the light aircraft spar under variable amplitude loading.

Simulation of Crack Propagation in Spar Using Extended Finite Element Method (XFEM)

The previous section showed it was possible, by using the finite element method and two-dimensional crack growth simulation in FRANC2D/L, to calculate the life of a spar exposed to a dynamic load so that it was approximate to the life obtained during the experiment; in the same way this section demonstrates that reliable life predictions can be made with crack growth simulation applied to three-dimensional spar geometry by using the extended finite element method (XFEM) [20].

XFEM is a relatively new method that is still not fully recognized and that still needs to prove its practical worth in order to be generally acknowledged. This means that the results obtained by using XFEM for a complex 3D geometry still cannot be regarded as reliable without experimental checks; therefore, the example of a spar life estimation given here should contribute to making a more objective judgment about XFEM usefulness.

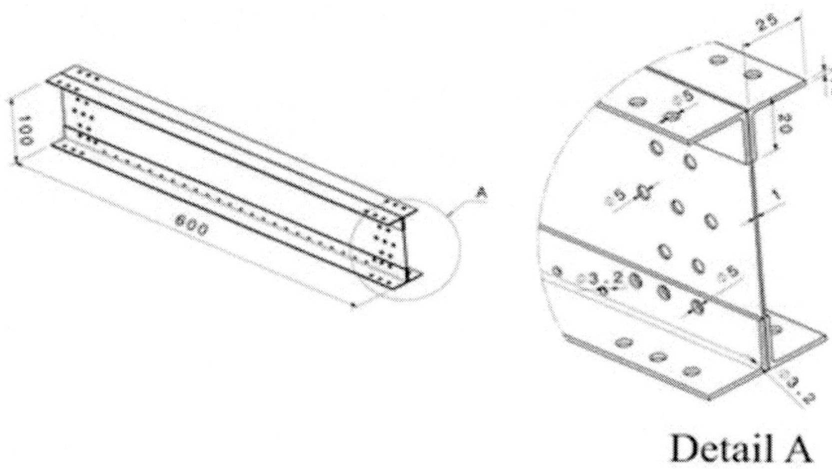

Figure 36. Dimensions and shape of an aluminum spar used in a 3D crack propagation simulation.

Figure 36 shows an aluminum spar modeled in CATIA v5 software. All dimensions were taken from the spars experimentally tested when exposed to various external loads by using the equipment described in the previous section. However, for the sake of a simplified numerical model and shorter estimation time it was necessary – before the virtual spar geometry was imported into Abaqus software – to remove all rivet holes along the four spar caps, as well as all thread connection holes on one spar side, which resulted in the geometry shape shown in Figure 37.

Figure 37 shows displacement (represented with arrows) applied at the end of the spar. The value of displacement used in the simulation was 3 mm which was the maximum displacement measured in the experiment with narrow band random loading. Holes for screws on the other side of the spar were used for constraining the model in the same manner as the real spar was constrained during the experiment.

Due to the specific geometry of the spar and the presence of a relatively large number of openings it was decided to create a finite element mesh consisting of tetrahedral elements.

Mesh used in calculations (shown in Figure 38) had 303,843 finite elements. It should be noted that first calculations were carried out on the finite element model with a larger number of elements (over 1 million), but the differences in results for both models were insignificant [21, 22]. In order to reduce time required for calculations, a simpler model was used.

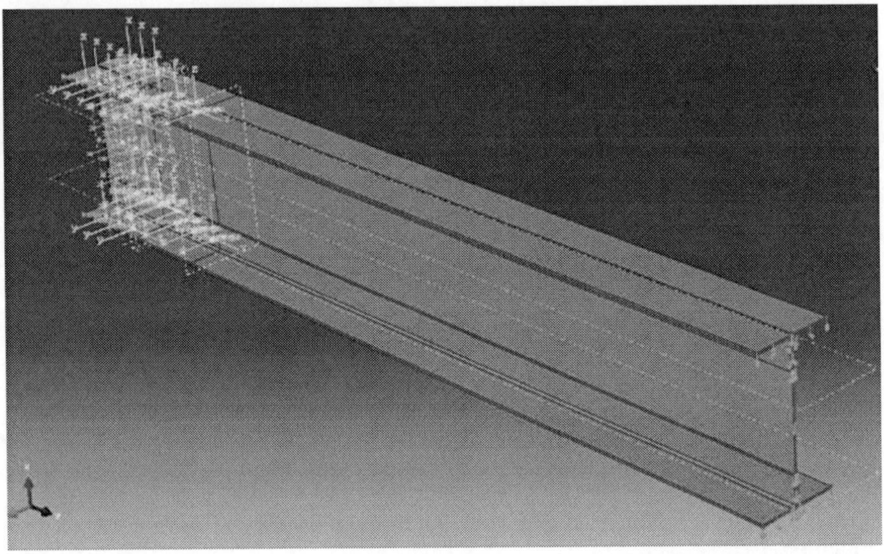

Figure 37. Model of spar with applied displacement at one end (Abaqus model).

Figure 38. Finite element mesh with initial penny-shape crack (dark circle).

As the first crack in the experiment appeared on the upper left spar cap (in the area close to the spar clamp), the mesh in the same area of the FE model was significantly refined in order to get more precise values of stress intensity factors (SIFs) at points on crack fronts.

In addition the initial crack was started at the edge of the left upper cap, which is not the exact position of crack initiation in the experiment with a narrow band random loading. The reason for choosing the edge for crack initiation was that during the experiments with different loading histories cracks always appeared in the same area but at different positions, and distances between these positions were about a few millimeters. This is reasonable, considering the stochastic nature of the fatigue phenomenon.

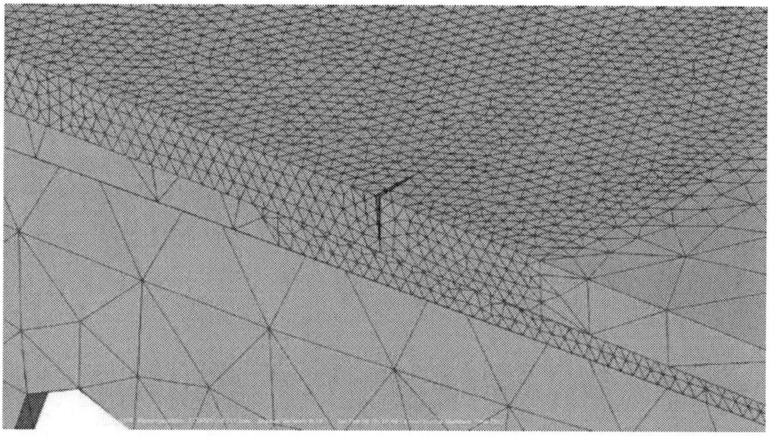

Figure 39. Spar cap after crack opening.

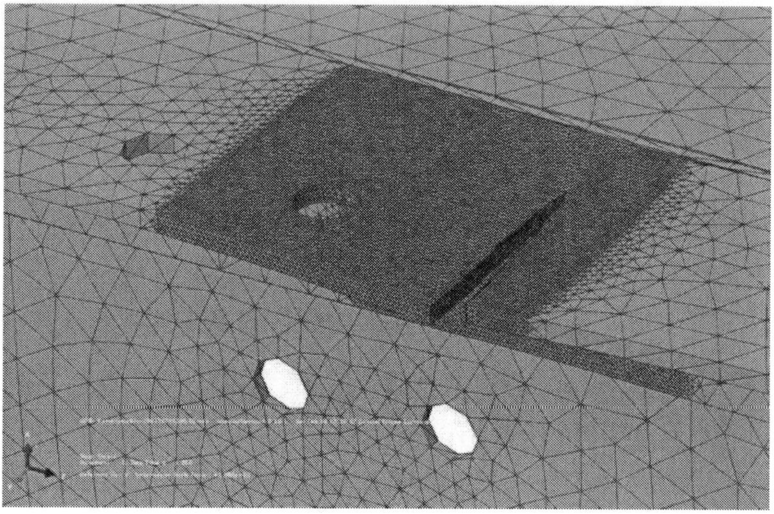

Figure 40. Crack shape after 17th step of propagation (top view).

Therefore, it was decided to provide more room for crack propagation in numerical simulation, in order to monitor and analyze changes in SIFs along the maximum possible path of the crack.

Initial crack length was 1 mm and, as can be seen in Figure 38, it didn't penetrate the entire thickness of the spar cap. Crack growth simulation and SIFs calculations were performed using MorfeoCrack add-in for Abaqus [23]. After initial calculations ("opening" of the crack), the finite element model changed its shape, as can be seen in Figure 39.

After opening, the crack was developed in stages of approximately 1 mm. Growth was not restricted to a single plane; instead the crack was allowed to grow in the direction predicted by MorfeoCrack for Abaqus. After seventeen steps of growth the FE model had a shape as shown in Figure 40, which shows the crack path on the horizontal wall of the left spar cap was not straight but curved, similar to that obtained in the experiment (Figures 13 and 16).

This path shape confirmed that boundary conditions and load applied as displacement were properly defined and that the numerical model simulates spar behavior sufficiently well.

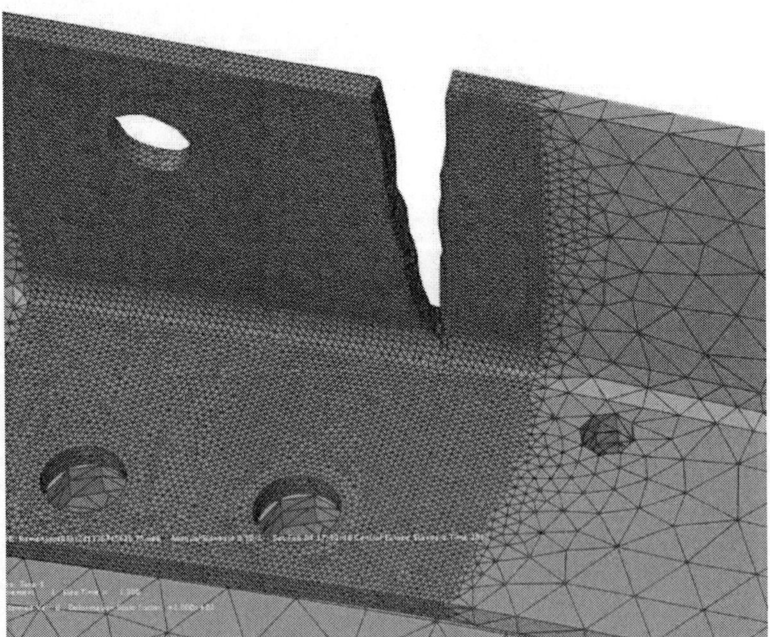

Figure 41. Crack shape after 24th step of propagation (bottom view).

Crack propagation was then continued. After 24 steps of growth (Figure 41) the crack reached the area of radius 3.2 mm between horizontal and vertical wall of the left upper cap (see Figure 36) and after the next few steps of propagation it left that area.

In this area contact between spar cap and spar wall begins and because cracks never appeared on the spar wall in the experiments, at the beginning of XFEM simulation the crack domain was limited only to the geometry of the left spar cap. Therefore, after leaving the area of radius 3.2 mm, the crack did not continue to spread along the spar wall (opposite to the experiment), but along the vertical wall of the spar cap which is in contact with the spar wall. After the 45th step of propagation (counting from the crack opening) the crack reached the edge of the vertical wall of spar cap and simulation stopped. This is because a total fracture of spar cap occurred, caused by the crack spreading along the total width of the spar cap walls (Figure 42).

Figure 42 shows that the crack path is not characterized by significant changes in a direction of propagation, but small deviations from the straight line. Comparing this virtual crack path obtained using XFEM and real crack paths obtained in experiments (Figure 13) it's obvious that paths do not have similar shapes.

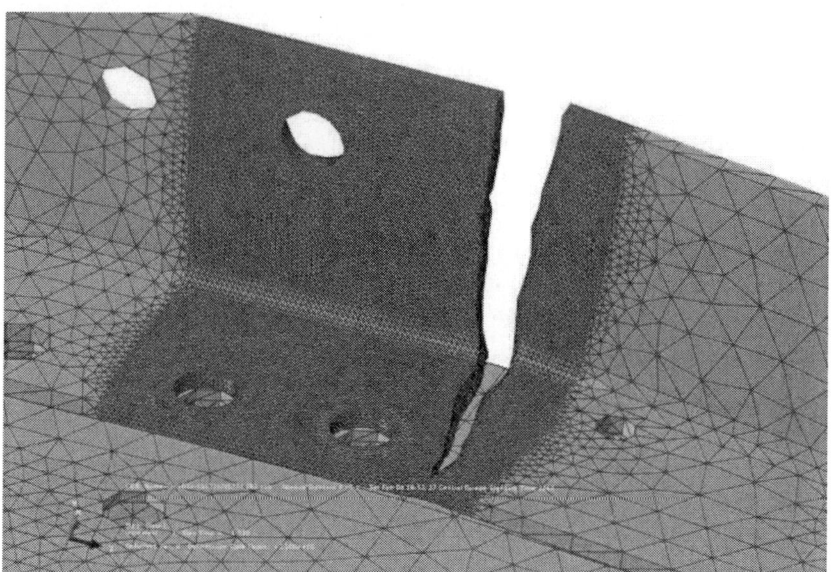

Figure 42. Crack shape after 45th step of propagation (bottom view).

In several experiments, after entering the area of radius 3.2 mm, cracks continued to develop along the area, at an approximate angle of 90° with respect to the path on the horizontal cap wall. After analysis of the cracks and fractures that occurred on spars during the experiment it was concluded that the caps made by bending thin 2024-T3 aluminum plates were exposed to residual tensile stresses in the area of radius 3.2 mm. After the cracks entered these areas the stress increased locally and that resulted in material destruction and crack propagation through the areas.

In experiments the cracks developed along these areas during several thousand cycles of external loading and since they were even some thirty millimeters long, the assumption of pre-stressed material in these areas was fully justified. Based on these studies it can be concluded that, during the manufacture of main structural components, it is necessary not only to choose high quality material, but also an adequate element-forming procedure and subsequent thermal treatment that would eliminate or considerably lower residual stress values.

Irrespective of differences between path shape obtained by 3D crack propagation simulation and actual crack shape obtained during the experiment with narrow-band random loading, it was determined to estimate the number of loading cycles which would propagate the initial crack from the edge of the horizontal wall along the whole width of the cap until its fracture, and to compare obtained value with the number of cycles during the experiment.

During each step of crack propagation values of stress intensity factors for various modes (I, II and III) were calculated at all points on generated crack fronts, as well as values of equivalent stress intensity factor K_{eq} [24].

Equivalent SIFs values obtained for 46 steps (crack opening and 45 propagations) are shown in Table 3, while graph in Figure 43 [25] shows a value of K_{eq} as a function of crack length.

Table 3 indicates that value of equivalent SIF increases until the twentieth step, when K_{eq}=790.442 MPa $mm^{0.5}$. After this step it starts to decline, which coincides with the crack entering the area between the horizontal and vertical wall of spar cap. The equivalent SIF value continues to decline until 26th step when crack leaves the area between horizontal and vertical cap wall, and then starts to grow again, until value K_{eq}=759.198 MPa $mm^{0.5}$ (step 29). After that, K_{eq} value oscillates in a relatively narrow range (from 634.529 to 768.188 MPa $mm^{0.5}$) until the final fracture that occurs in 46th step.

Once the SIF values for each step had been calculated it was possible to evaluate the number of loading cycles which would propagate a crack to a given length.

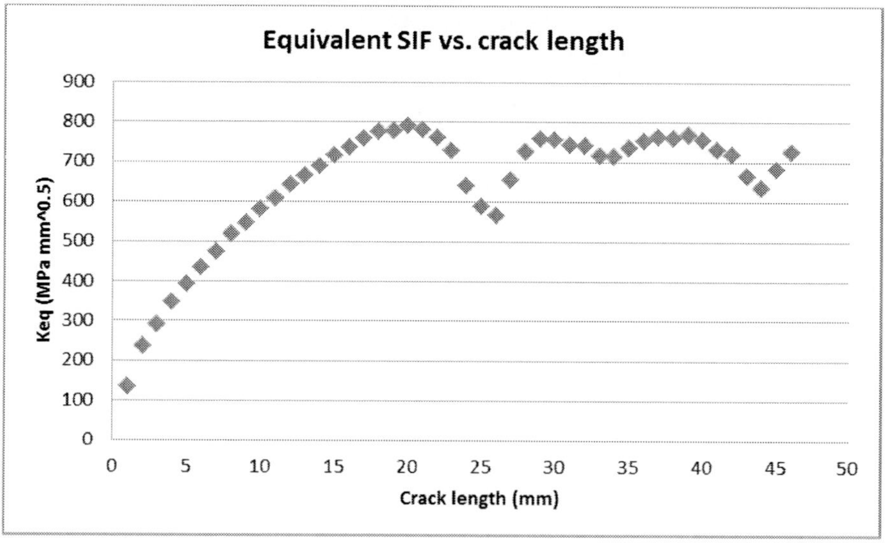

Figure 43. Values of equivalent stress intensity factor (K_{eq}) as a function of crack length.

Table 3. Values of equivalent SIF (min, max and mean for each crack front) obtained in XFEM simulation of crack growth

	Equivalent SIF (Keq) (MPa $mm^{0,5}$)		
Step	Max. value	Min. value	Mean value
1	138.678	134.545	136.025
2	253.886	226.932	236.461
3	296.453	288.411	291.978
4	355.690	339.466	347.197
5	395.794	387.573	392.023
6	450.694	426.088	435.673
7	478.861	467.768	474.237
8	526.203	513.699	519.532
9	556.621	537.933	546.987
10	585.258	576.896	580.596
11	629.285	592.731	608.985
12	676.938	603.792	642.468
13	677.887	653.492	665.442
14	697.955	681.625	689.093
15	727.070	700.509	716.752
16	748.252	725.788	737.541

Table 3. (Continued)

	Equivalent SIF (Keq) (MPa $mm^{0,5}$)		
Step	Max. value	Min. value	Mean value
17	769.358	738.295	758.608
18	792.089	768.371	775.517
19	793.738	755.383	779.585
20	807.831	776.104	790.442
21	782.835	778.856	780.780
22	785.557	734.591	762.366
23	742.274	697.416	729.154
24	659.412	620.302	640.564
25	648.635	537.646	587.540
26	620.583	522.437	566.972
27	869.697	485.310	654.067
28	850.326	581.597	725.730
29	810.775	703.084	759.198
30	771.231	742.939	755.954
31	757.192	732.311	744.414
32	747.939	736.862	741.725
33	728.988	700.861	717.330
34	724.062	701.643	715.498
35	784.156	697.312	737.215
36	767.658	743.145	752.754
37	781.965	734.106	763.236
38	777.65	742.239	760.791
39	800.329	718.432	768.188
40	767.17	751.343	757.403
41	754.186	705.533	731.028
42	742.684	697.336	718.670
43	697.121	622.947	664.065
44	672.574	585.665	634.529
45	743.352	643.458	682.498
46	739.827	721.688	727.443

The fatigue crack growth model incorporated in MorfeoCrack for Abaqus is based on Paris' Law and it requires values for the coefficients C and m for aluminum alloy 2024-T3 to be entered, as well as the stress ratio R (in this case R=0.15).

The values obtained in MorfeoCrack for Abaqus (Figure 44) show that the initial crack (length 1 mm) will extend to 2 mm after approximately 27,000 cycles of applied displacement, while it will grow up to 3 mm after another 5,850 cycles. The crack will enter the area between the horizontal and vertical walls of the spar cap after approximately 45,000 cycles from initiation. It will take another 5,000 cycles from the instant when the crack leaves this area (step 26) to the final fracture of upper left spar cap (step 46). The total number of cycles that propagate the initial crack to its final length (as obtained in MorfeoCrack for Abaqus) is 50,743. If we add this value to the number of cycles required to initiate the crack on the spar cap (7,944 cycles, value obtained in Ansys) we get estimated fatigue life N = 58,687 cycles.

The differences between the paths of actual and virtual cracks, as well as between the starting and ending points between which cracks propagated on the actual and virtual spars lead to the conclusion that it makes sense to compare numbers of cycles (i.e. actual and estimated fatigue life) only in situations when cracks have similar behavior, where there are no residual stresses and when a virtual crack is initiated at the point where it actually appeared.

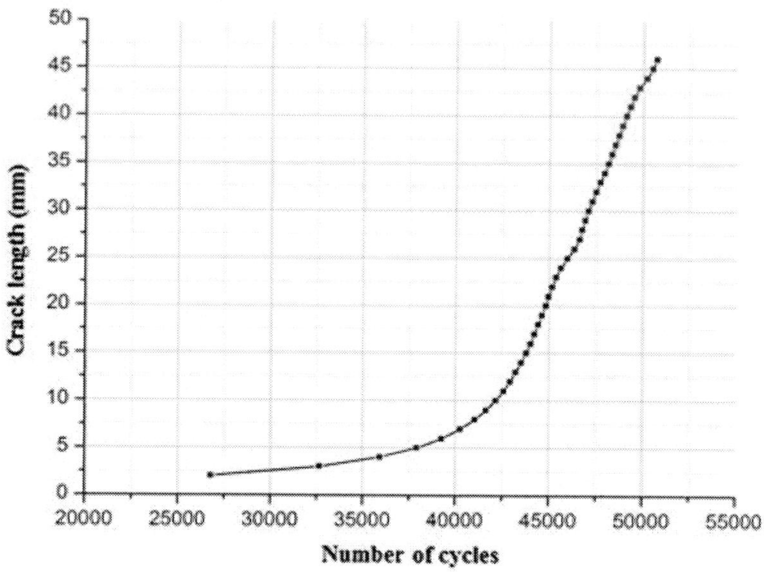

Figure 44. Crack length as a function of the number of cycles (applied displacements 3 mm).

In order to confirm this, several additional experiments were conducted on spars which were, unlike those previously used, made of enhanced aluminum alloy 2024-T3 ($C = 6.106\times10^{-11}$, $n= 2.60$); also, after cold forming, spar caps were softly annealed and exposed to the stabilization process [4] with the aim of reducing residual stresses.

Conditions in which the experiments with enhanced alloy spars were conducted, including applied loading spectra, were identical to those described in the previous section in order to compare new with previously obtained results. During the new experiments two things were immediately noticed: a larger number of external loading cycles was necessary for propagating the cracks along the lengths from previous experiments; the cracks had different propagation paths along the caps.

Figure 45 shows that the crack on one of new spars propagated along the path is very similar to the one obtained by using Abaqus simulation (Figure 46) and that it did not propagate along the area of radius 3.2 mm.

Figure 45. View of the crack on the spar made of enhanced alloy 2024-T3.

This fact additionally confirms the conclusion that the main cause for previous crack propagation along this area was the residual stress that was almost completely eliminated on the new cap by using the stabilization process. Besides, before the testing took place, a small notch was cut on the spar shown in Figure 45 at the place where the initial crack was set in Abaqus so that the actual crack during the experiment could develop from the same place as the virtual one during the simulation.

In this case it was not necessary to determine, by using Ansys software, the number of cycles until the appearance of the initial crack since the crack was already present at the spar cap.

The crack shown in Figure 45 produced this shape after 64,148 cycles of applied load, which was considerably less than 76,758 cycles predicted for the enhanced alloy spar by modified Paris model integrated with MorfeoCrack for Abaqus. Since it was obvious that the estimation based on a modified Paris model might endanger structural integrity, the decision was made to estimate the life by using a Forman-Mettu NASGRO equation (1).

Stress intensity factor values previously obtained in Abaqus (Table 3) were used in F2D/LCG software – in the way described in the previous section – so that the estimated fatigue life of enhanced alloy 2024-T3 could be obtained. Figure 43 shows considerable K_{eq} variations for lengths larger than 25 mm since then the crack propagated from horizontal to vertical spar cap wall and changed propagation direction.

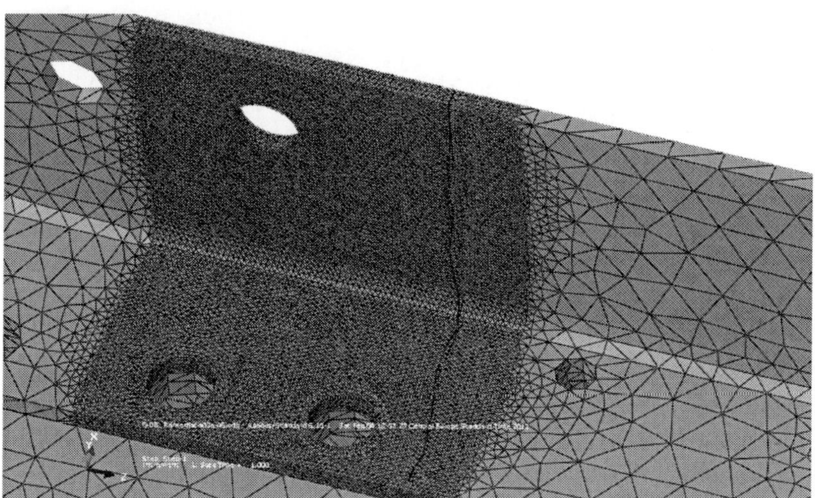

Figure 46. View of the crack in simulation after 45 propagation steps (1 mm each).

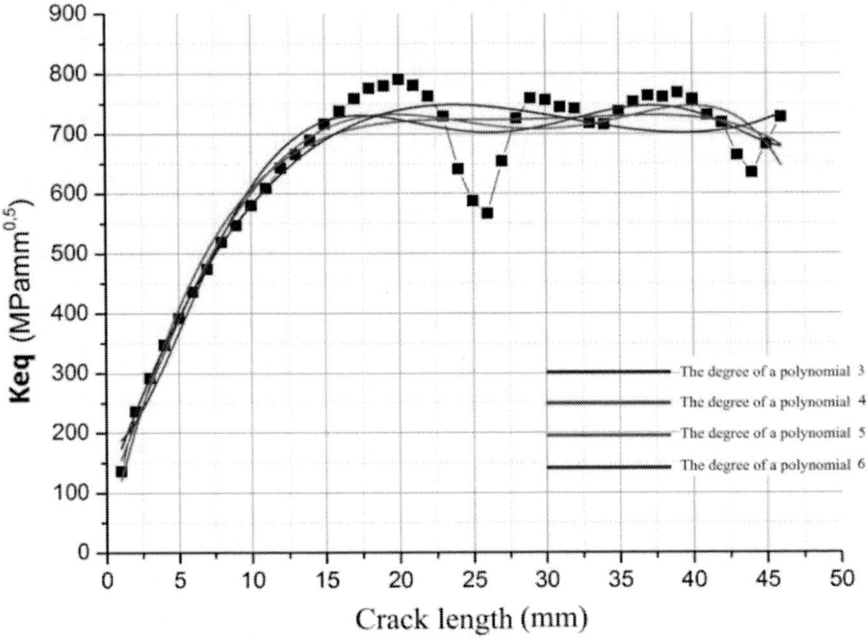

Figure 47. Curve of equivalent stress intensity factor change approximated with higher degree polynomial functions.

Since F2D/LCG software uses the graph shown in Figure 43 in order to find a trend line equation (necessary for integration of the NASGRO equation), an inadequately chosen trend line would lead to an insufficiently precise life estimation. It is obvious that the line shown in Figure 43 cannot be precisely defined with any polynomial function (Figure 47); since the shape of fifth-degree polynomial function was the most similar to K_{eq} function, that was the trend line for which the fatigue life was calculated. Obtained value of cycle numbers (N = 61,991) was the most approximate to the experimentally obtained value. For more precise fatigue life predictions, it is obvious that the trend line needs to be as close to the K_{eq} change graph as possible; if this is not possible to achieve several lines may be used [21], i.e. equivalent stress intensity factor curve should be approximated in segments. In this case it was achieved with two trend lines (one to represent part of the curve up to the crack length of 26 mm, and the second to represent the remaining part); after, by the addition of the fatigue life value for the first (56,812 cycles) and the second segment (5,722 cycles), a total life N = 62,534 was obtained, and this value was very near to the experimentally estimated life (N = 64,148 cycles).

Conclusion

Accurate matching of spar cap crack fatigue life values calculated by using XFEM and actual fatigue life values, as well as similarities in crack paths, enable us to draw several conclusions:

1. The extended finite element method gives good predictions related to stress intensity factor values for complex load-carrying structures made up of several elements.
2. When estimating the fatigue life of complex load-carrying structures it is reasonable to use equivalent stress intensity factors (K_{eq}) taking into account values of all three modes (K_I, K_{II}, K_{III}).
3. When estimating the fatigue life of load-carrying structures it is very important to choose the appropriate polynomial function that will be used during integration of the da/dN expression, for approximation of K_{eq} function in relation to crack length.
4. Regarding more complex curve shapes describing K_{eq} change in relation to crack length (as in case shown in Figure 43), it is recommended using several polynomial functions for more precise approximation of certain curve segments. Afterwards, total fatigue life is estimated by adding the estimated lives for all individual curve segments.
5. Fatigue characteristics of used materials, thermal treatment and stabilization after forming of assembly elements have considerable influence on the direction in which cracks spread through the structure as well as on fatigue life values. By using XFEM and FEM it is possible to predict the most probable directions in which cracks will develop through the structure; if, in actual case, it propagates in some other directions, the reasons leading to this should be determined.
6. Experimental tests should still be conducted whenever it is possible and economically reasonable, but well defined and verified numerical model can give fatigue life values approximate to the ones obtained during the experiments. Besides, 3D models can be used for researching the fatigue behavior of load-carrying structures made up of different materials, which can – based on comparing obtained results and detailed economical analysis – determine the alloy providing the longest possible fatigue life within the expected service environment [26].

References

[1] Boyer, H. E. (editor), *Atlas of Fatigue Curves,* American Society for Metals, February 1986.

[2] Rasuo, B., Experimental techniques for evaluation of fatigue characteristics of laminated constructions from composite materials: full-scale testing of the helicopter rotor blades, *J. Testing Evaluation ASTM* 2011; 39:237-242.

[3] Hiscocks, R., *Design of Light Aircraft*, DARcorporation, ASIN: B000 ZMF1V6, 2002.

[4] Rasuo, B., *Aircraft production technology* [in Serbian], Faculty of Mechanical Engineering, University of Belgrade, 1995.

[5] Kvačkaj, T., Bidulský, R. (editors), *Aluminum Alloys, Theory and Applications,* InTech, Croatia, 2011.

[6] Grbovic, A., Rasuo, B., FEM based fatigue crack growth predictions for spar of light aircraft under variable amplitude loading, *Engineering Failure Analysis*, Volume 26, p. 50-64, 2012.

[7] Tovo, R., Cycle distribution and fatigue damage under broad-band random loading, *Int. J. Fatigue* 24(11):1137-1147, 2002.

[8] Nagode, M., Fajdiga, M., On a new method for prediction of the scatter of loading spectra, *Int. J. Fatigue* 20(4):271-277, 1998.

[9] Klemenc, J., Fajdiga, M., Description of statistical dependencies of parameters of random load states (dependency of random load parameters), *Int. J. Fatigue* 22(5):357-367, 2000.

[10] Niu, M., *Airframe Structural Design,* Technical Book Co, Hong Kong, 1988.

[11] *ASM Handbook*, Volume 19, Fatigue and Fracture, 1997.

[12] Xiong, J. J., Shenoi, R. A., *Fatigue and Fracture Reliability Engineering*, Springer-Verlag London Limited, 2011.

[13] www.ansys.com.

[14] www.cfg.cornell.edu/software/software.htm.

[15] Mitchell, M., Landgraf, R. (editors), *Advances in fatigue lifetime predictive techniques,* 3rd Volume, ASTM, 1996.

[16] Bower, A., *Applied Mechanics of Solids*, CRC Press, New York, 2009.

[17] Liu, A., *Mechanics and Mechanisms of Fracture: An Introduction*, ASM International, Ohio, 2005.

[18] Forman, R. G., Mettu, S. R., In: *ASTM STP 1131* (edited by Ernst, H. A. et al.), p. 519-546, 1992.

[19] McKeighan, P., Ranganathan, N. (editors), Fatigue Testing and Analysis Under Variable Amplitude Loading Conditions, *ASTM*, p. 245, Mayfield 2005.

[20] Rasuo, B., Grbovic, A., Petrasinovic, D., Investigation of Fatigue Life of 2024-T3 Aluminum Spar Using Extended Finite Element Method (XFEM), *SAE Int. J. Aerospace* 6(2):408:416, 2013.

[21] Grbovic, A., *"Investigation of fatigue life in superalloys structural components", Ph.D. Thesis*, University of Belgrade, Faculty of Mechanical Engineering, Belgrade, 2012.

[22] Petrasinovic, D., *"Fatigue life analysis of metal wing structural parts", Ph.D. Thesis,* University of Belgrade, Faculty of Mechanical Engineering, Belgrade, 2012.

[23] *MorfeoCrack for Abaqus,* GeonX, Belgium, http://www.geonx.com/, accessed on 2013-03-29.

[24] Berkovic, M., "Determination of Stress Intensity Factors Using Finite Element Method", *Structural integrity and life*, vol. 4, No 2, p. 57-62, 2004.

[25] Petrasinovic, D., Rasuo, B., Petrasinovic, N., "Extended finite element method (XFEM) applied to aircraft duralumin spar fatigue life estimation", *Technical Gazette*, Vol. 19, p. 557-562, 2012.

[26] Grbovic, A., Rasuo, B., Vidanovic, N., Peric, M., Simulation of crack propagation in titanium mini dental implants (MDI). *FME Transactions* 2011; 39:165-170.

In: Recent Trends in Fatigue Design
Editor: Ricardo Branco
ISBN: 978-1-61668-410-5

Chapter 3

CERTAIN ASPECTS OF NOTCH FATIGUE

***N. Ranganathan*[*] *and R. Leroy*[#]**
Laboratoire de Mécanique et Rhéologie, Université de Tours
Polytech Tours, Tours, France

ABSTRACT

This paper treats certain aspects of notch fatigue. After a short survey of different issues, a short crack model for crack growth kinetics at a notch is presented. In this model, a small crack is assumed to exist at a notch and the crack initiation life is in fact the propagation life for this crack to attain a detectable size. This aspect is verified using specially designed tests in an aluminum alloy. Finally, comparing the short crack approach to other conventional methods shows that this model gives acceptable prediction for crack initiation life for constant and variable amplitude loading in fatigue.

Keywords: Notch effect, fatigue life estimation, short crack effect, modeling

---

[*]Professor, Head, and [#]Associate Professor, Laboratoire de Mécanique et Rhéologie, Université de Tours, Polytech Tours, 7 avenue Marcel Dassault, 37200 Tours, France. Email: ranganathan@univ-tours.fr

INTRODUCTION

Prediction crack initiation life at a notch is very critical, as notches act as stress concentrators. This phenomenon is critical for aircraft components as they are designed with very small tolerances. One of the most often cited failure due to notch effect is that of the Comet aircrafts in the 1950s [1].

In this paper, firstly the common approaches for notch fatigue evaluation are succinctly presented.

This is followed with the presentation of the short crack approach and a flow chart of a model is presented.

Specially designed tests to verify the short crack approach are discussed. These results show that the model assumptions are valid.

BACKGROUND

Conventional Approaches

Firstly, the notch effect is illustrated in terms of stresses in the vicinity of a circular notch. We shall call, s, as the remotely applied stress and σ, the stress in the vicinity of the notch.

The stress concentration factor is defined as

$$k_t = \sigma_y\,(\max)/s \tag{1}$$

where σ_y (max) is the maximum local stress, at the notch tip along the y direction (see figure 1)

Note that the local stresses fall of rapidly as the distance from the notch tip increases, according to the following equation:

$$\frac{\sigma_y}{s} = 1 + 0.5(r/x)^2 + 1.5\,(r/x)^4 \tag{2}$$

Where r is the radius of the hole and x, the distance from the center of the hole

Along the x direction stresses vary according to

$$\frac{\sigma_x}{s} = 1.5(r/x)^2 - 1.5\,(r/x)^4 \tag{3}$$

In fact we have a biaxial stress situation due to the notch effect. The k_t value for this geometry is found to be 3. Stress concentration factors for can be found in [2].

Under fatigue loading , the presence of a notch decreases the endurance limit and also life in the finite life regime is decreased. These aspects are now presented. The decrease in endurance limit due to a notch is taken into account in terms of the fatigue notch factor $k_{f.,}$ defined as:

$$k_f = \frac{\sigma_d}{\sigma_{dn}} \tag{4}$$

Where σ_d is the endurance limit of the material and σ_{dn} the endurance limit in the presence of a notch.

k_f anf k_t are related according to

$$q = \frac{k_f - 1}{k_t - 1}. \tag{5}$$

Here q is the notch sensitivity factor.

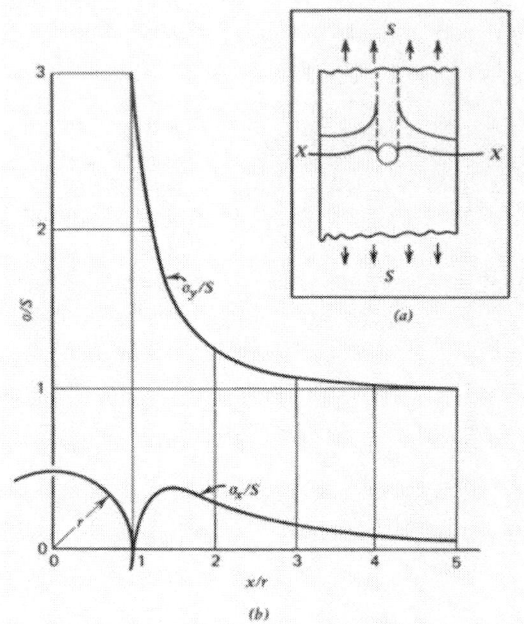

Figure 1. Stress distribution at circular notch.

Empirical formulae exist to estimate the factor q, that depends upon the material. The most famous expression is that of Peterson [3],

$$q = \frac{1}{1+\frac{a}{r}} \tag{6}$$

Where r is the notch root radius and a, a material constant. The value of “a” for aluminum is 0.635mm.

In the case of finite life regime, the local strain amplitude at the notch tip can be evaluated using either the Neuber’s hypothesis or the Glinka’s energy based approach [4]. These methods are explained below:

In the Neuber’s approach, two factors are introduced: the stress intensity factor, k_σ and k_ε, the strain intensity factor, that correlate remote and local stress-strain fields,

$$k_\sigma = \frac{\sigma}{s} \tag{7}$$

$$k_\varepsilon = \frac{\epsilon}{e} \tag{8}$$

In these equations, σ and ε are local stress and strain at the notch tip, while s and e are remote stress and strain values.

So long as the local stresses are in the elastic regime, all the concentration factors are equal, i.e.

$$k_\sigma = k_\varepsilon = k_t \tag{9}$$

When the local stresses reach the material yield stress, the local stress values increase at a much lower rate than in the elastic regime, while the local strains increase much faster than in the elastic regime, i.e.

$$k_\sigma < k_t < k_\varepsilon \tag{10}$$

According to Neuber's hypothesis, the following hyperbolic equation, represents the elastic-plastic regime:

$$k_\sigma \, . \, k_\varepsilon = k_t^2 \tag{11}$$

This equation can be rewritten in the case of fatigue loading,

$$\frac{\Delta S^2 {k_t}^2}{E} = \Delta\sigma . \Delta\varepsilon \quad (12)$$

Here, $\Delta\sigma$ and $\Delta\varepsilon$ are local stress and strain amplitudes while Δs is the remotely applied stress amplitude. E is the Young's modulus of the material

The above equation is combined with the material constitutive law, using the Ramberg -Osgood representation

$$\frac{\Delta\varepsilon}{2} = \frac{\Delta\sigma}{2E} + \left(\frac{\Delta\sigma}{2\,K'}\right)^{\left(\frac{1}{n'}\right)} \quad (13)$$

Here K' is the cyclic strength coefficient and n' the cyclic strain hardening exponent.

In the Glinka's energy based approach, the local strains and stresses are determined from an energy equivalence hypothesis, leading to the following equation

$$\frac{\Delta S^2 {k_t}^2}{E} = \frac{\Delta\sigma^2}{E} + \left(\frac{2\Delta\sigma}{n'+1}\right)\left(\frac{\Delta\sigma}{2K'}\right)^{\left(\frac{1}{n'}\right)} \quad (14)$$

For the Neuber approach, equations 12 and 13 are numerically solved to determine the local strain or stress amplitudes, while for the Glinka Model, equations 13 and 14 are solved numerically. Comparison between both the methods show that the estimated strains by the Glinka Model are slightly lower than those obtained using the Neuber approach [4].

Having determined the local strain amplitude, the corresponding fatigue life can be determined using the conventional material strain life relationship:

$$\frac{\Delta\varepsilon}{2} = \frac{\sigma_{f'}}{E}{N_f}^b + \varepsilon_{f'}{N_f}^c \quad (15)$$

Here $\sigma_{f'}$ and $\varepsilon_{f'}$ are material constants, determined from constant amplitude tests and b and c are Basquin's and Coffin-Manson exponents [5].

N_f, determined from the above equation, represents in fact the crack initiation life at the notch tip. However, the size of the crack is unknown.

To overcome this shortcoming, a short crack model has been proposed. This model is now presented in detail.

SHORT CRACK MODEL

As opposed to the conventional models presented above, in the short crack model a defect of a small size is assumed present at the notch tip. This defect can be an inclusion, an intermetallic precipitate or a void. This defect is assimilated to a short crack. This crack would grow under fatigue loading, following short crack kinetics. In this model, crack initiation is assumed to occur, when this short crack attains a length detectable by modern NDT techniques. Its size is around 0.5mm.

One of the difficulties of this model is that the crack geometry is unknown. In our model, we use the edge crack, a semi elliptical embedded crack or a through-thickness crack. The estimated stress intensity factors for the three geometries are compared in figure 2.

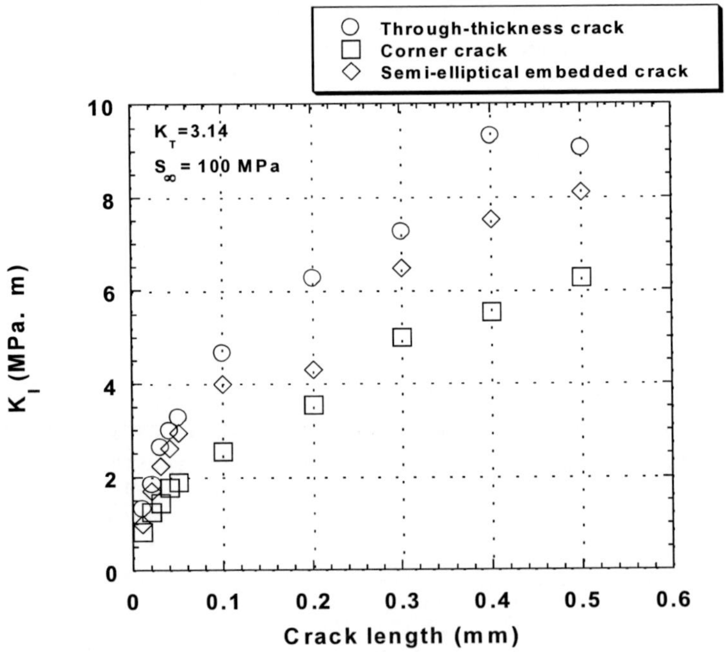

Figure 2. Stress intensity factors for three crack geometries.

The estimations are made using the Kujawski approximation or the model presented by Newman and Raju [6,7]. A k_t value of 3.14 was considered and the remote stress was 100 MPa. The local stress intensity factors are determined after numerically solving for the notch tip stress level, using the Glinka approach.

It can be seen here that the stress intensity factor, K, depends upon the crack geometry in the following order:

$$K_{\text{through-thickness}} > K_{\text{corner crack}} > K_{\text{elliptic embedded crack}} \quad (16)$$

This would mean that the crack growth kinetics should also depend upon the crack geometry. A home built computer algorithm has been developed, that permits the determination of the crack initiation life using the conventional approach and the short crack model. The local strain approach uses the above mentioned calculation methods (Neuber or Glinka) to evaluate the local strain history, and a simple linear damage summation law is assumed. The damage due to each elementary cycle is taken from normalized tests on smooth specimens submitted to fully reversed strain amplitude (R_ε=-1). It is represented by an elementary damage variable $D_i = \frac{1}{N_i}$ (N_i being the failure life obtained for a smooth specimen submitted to the same strain amplitude). The crack is assumed to be initiated when the sum of each elementary damage $\sum_{i=1}^{N} D_i = D$ equals 1. This algorithm is presented in figure 3. In the software, the following mean stress corrections are available: Morrow, Manson-Halford and Smith-Watson-Topper. The Manson-Coffin correction for the strain-life curve is available too. Details of these corrections are given in [8]. These corrections are applied only for strain amplitudes lower than the elastic-plastic transition limit.

In this paper, firstly, certain tests concerning the verification of the short crack model are presented. Then results of notch fatigue under constant and variable amplitude loading are compared with model prediction.

Experimental Details

The present study was conducted on two aluminum alloys, the 7449 alloy in the T7951 condition and the 2024 alloy in the T351 condition. The nominal composition and mechanical properties are given in tables 1and 2 respectively.

Table 1. Composition (in %wt.) of the studied alloys (Al not indicated)

Alloy	Zn	Mg	Cu	Cr	Fe	Si	Mn	Ti
7449	7.5-8.7	1.8-2.7	1.4-2.1	-	0.15 max	0.12 max	0.2 max	0.25
2024	0.04	1.5	4.46	0.01	0.22	0.1	0.66	0.02

Table 2. Nominal Mechanical Properties

Alloy	Young's Modulus (MPa)	0.2% YS (MPa)	UTS (MPa)	Elongation (%)
7449 T7951	71,500	598	623	8
2024 T351	73,000	300	500	16

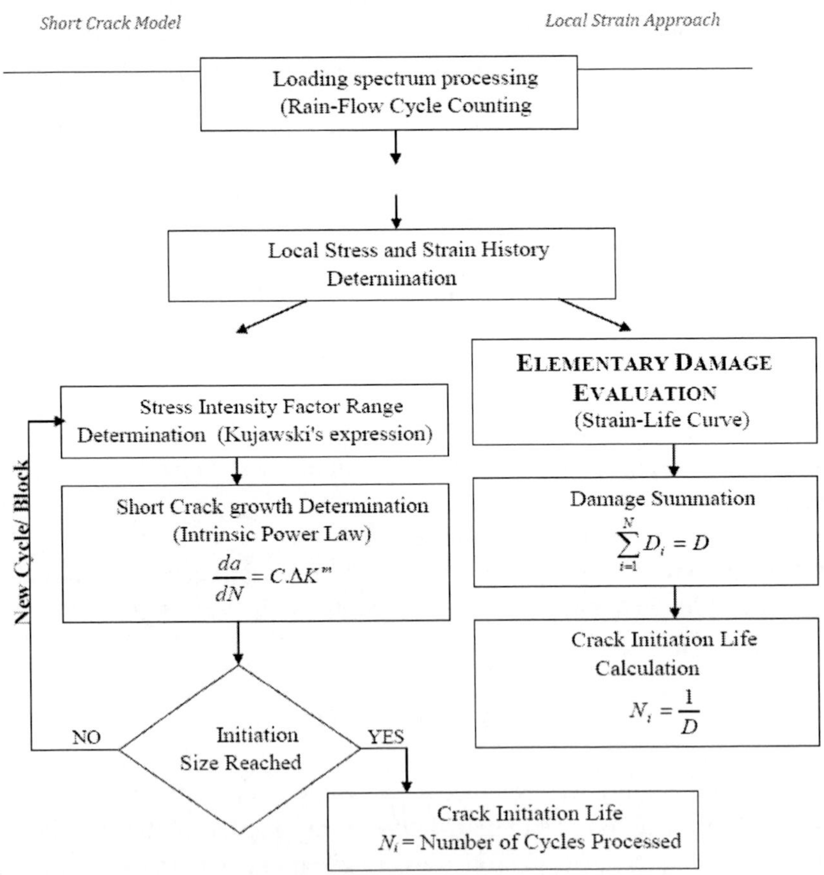

Figure 3. Flow chart of the proposed algorithm.

In the 7449 alloy, specially designed marker bands tests were carried out using specimens with a k_t of 3.41. The specimen geometry is given in figure4 and the marker band load sequence in figure5.

The marker band loading consists of 10000 cycles at a baseline load, at a load ratio of 0.1, interspersed with 1000 marker cycles at a load ratio of 0.8. Within the marker cycles an over load under-load combination was added. This sequence was verified to leave well defined marker bands on the fracture surface (seen under a scanning electron microscope), without modifying the crack growth kinetics [9].

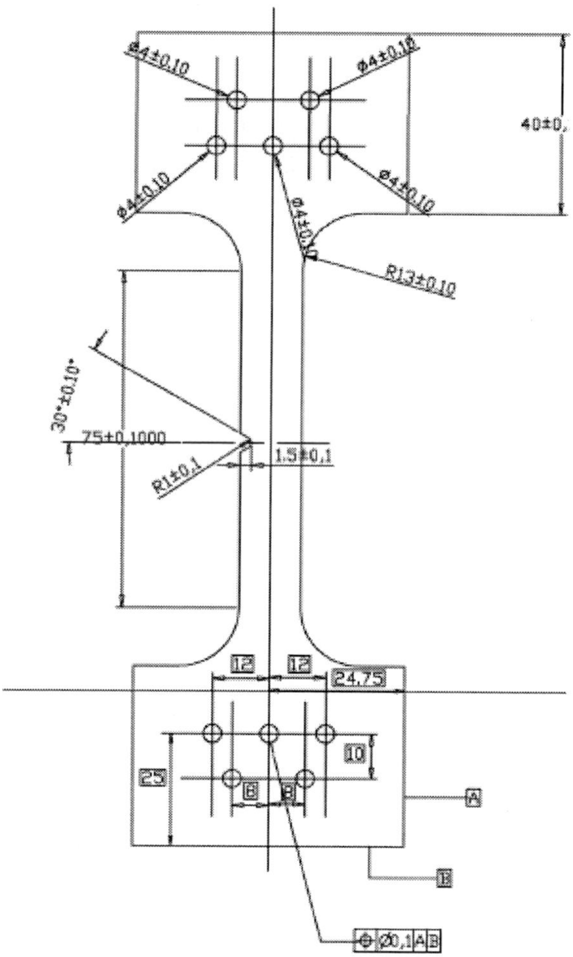

Figure 4. Specimen for the 7449 alloy.

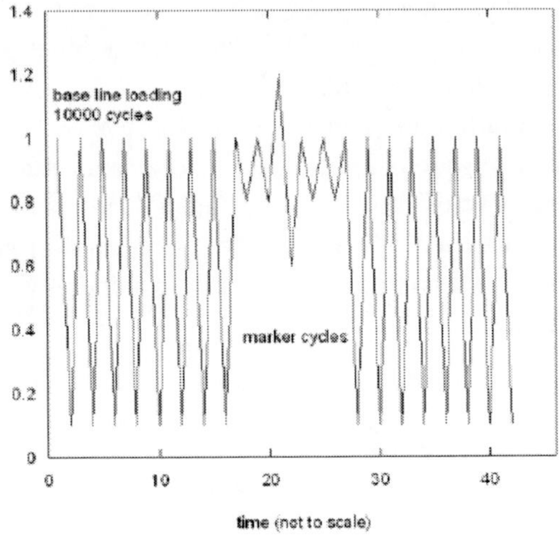

Figure 5. Marker band loads.

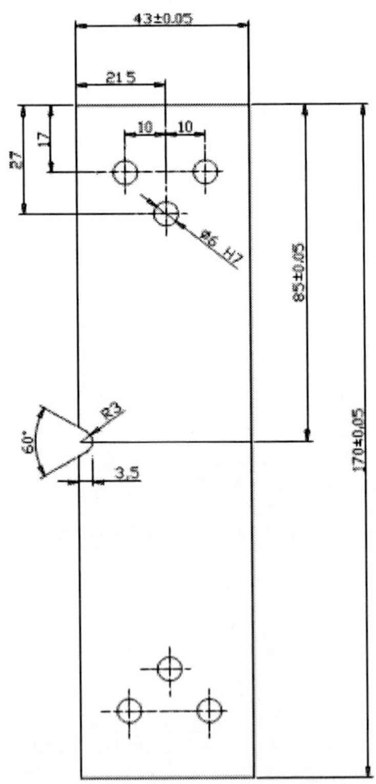

Figure 6. Notched specimen for 2024 alloy.

In the 2024 alloy, another geometry, with a k_t of 3.14 was used, shown in figure 6. In this alloy constant amplitude and variable amplitude tests were carried out. Details can be found in [10].

RESULTS

Results of marker band tests in the 7449 alloy are first presented. These tests were carried out at a load ratio of 0.1. Two stress levels of 120 and 140 MPa were considered. The total lives to failure for the two tests were 348000 and 193000 cycles respectively. Figure 7 shows marker bands at the crack initiation site and figure 8 at 1mm from the notch.

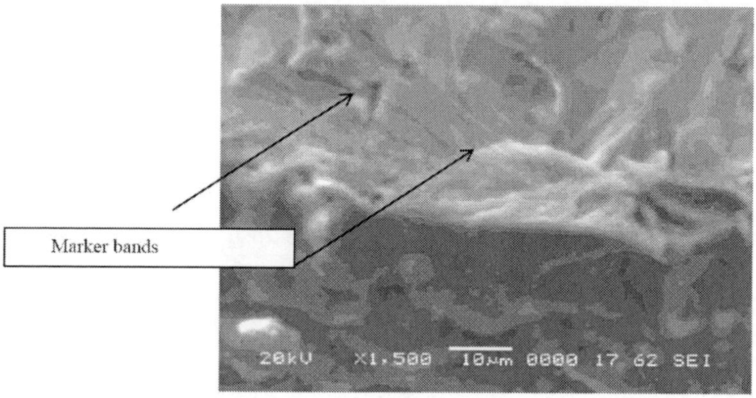

Figure 7. First marker bands around an inclusion.

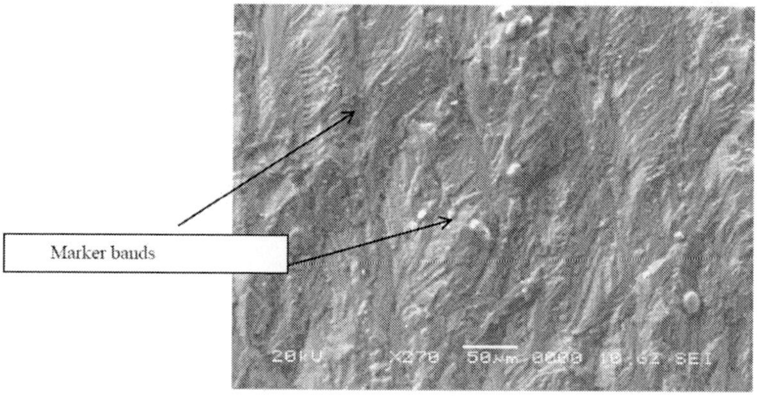

Figure 8. Marker bands 2 mm from the notch.

Marker bands are clearly visible. Knowing that the number of baseline load cycles is 10000, the average crack growth rate between marker bands can be estimated by;

$$da/dN = \Delta_{mb}/10000 \quad (17)$$

Where Δ_{mb} is the average distance between the bands; For each case, three measurements between the bands were made to get the average spacing.

Table3 presents the analysis for the test at 140 MPa.

Table 3. Marker band analysis

distance from notch a(mean)	Δmb (mm)	da/dn(μm/cycle)	N(cycles)
160.7	164	16.4	4799
377.2	269	26.9	6097
603.7	184	18.4	14620
858.7	326	32.6	5644
1181.7	320	32	10188
1468.2	253	25.3	12648
1672.7	156	15.6	16218
1830.7	160	16	9750
2025.2	229	22.9	6987
2247.2	215	21.5	10651
2560.7	412	41.2	5218
2971.7	410	41	10049
3317.2	281	28.1	14591
3578.2	241	24.1	11660
3904.2	411	41.1	5864
4324.7	430	43	9558
4738.7	398	39.8	10804
5134.2	393	39.3	10127
5557.2	453	45.3	8675
6433.7	1300	130	3485

From the average crack growth rate, the number of cycles in each interval is determined. The sum of all this data gives the total crack propagation life N_p, which is 187600 cycles. The crack initiation life can be obtained by

$$N_i = N_f - N_p \tag{18}$$

The results are given in the following table.

Table 4. Initiation and propagation lives

Stress amplitude (Mpa)	Nf (Cycles)	Np (Cycles)	Ni (Cycles)	Initiation life %
120	3.48E+05	2.65E+05	8.30E+04	23.9
140	1.93E+05	1.87E+05	6.00E+03	3.1

It can be seen here that the initiation life is about 24% for the first test while it is only 3% for the second one. These results show that crack initiation lives can be neglected at least for lives up to 193000 cycles.

2024 T351 Alloy

In this alloy, we used a specimen with a k_t of 3.14 (fig.6). Firstly constant amplitude tests were carried out covering a range from 120 to 260 MPa. For these tests, growth kinetics was followed by metallurgical replicas on the polished specimen surface. These replicas were taken at regular intervals, at the maximum load (so that the crack, if any, is fully open). The crack imprint on the replica was measured using a metallurgical microscope.

Comparison between the measured and estimated crack growth kinetics is carried out in fig.9. For the estimations, we used both the corner crack and the though thickness crack configurations. It can be seen that the measured crack growth kinetics can be accurately predicted considering appropriate crack geometries.

The next figure compares the predicted and estimated crack initiation lives for the series of tests. The initiation crack size is taken as 500µm. This covers all the tests at a load ratio of 0.1. It can be confirmed that the short crack model correctly predicts the evolution of crack initiation lives and all the experimental data points (except one) fall within the scatter band determined by the two crack configurations.

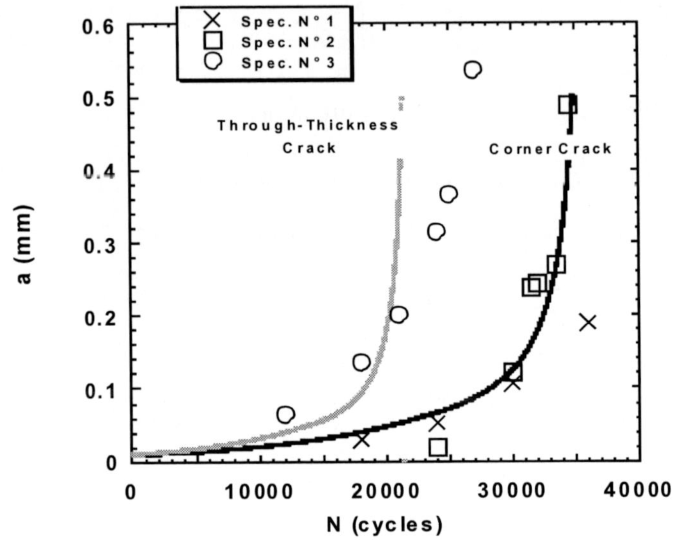

Figure 9. Measured and estimated crack length evolutions for three tests.

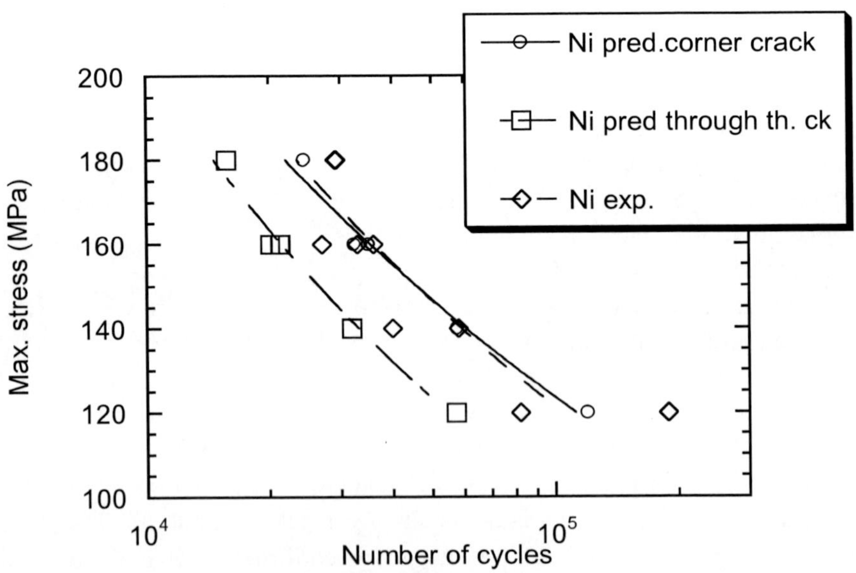

Figure 10. Measured and estimated lives for two crack geometries.

Next, we analyze data in the literature [11] where crack initiation lives were studied in the same alloy under an aircraft spectrum loading. A typical flight type is shown in figure 11.

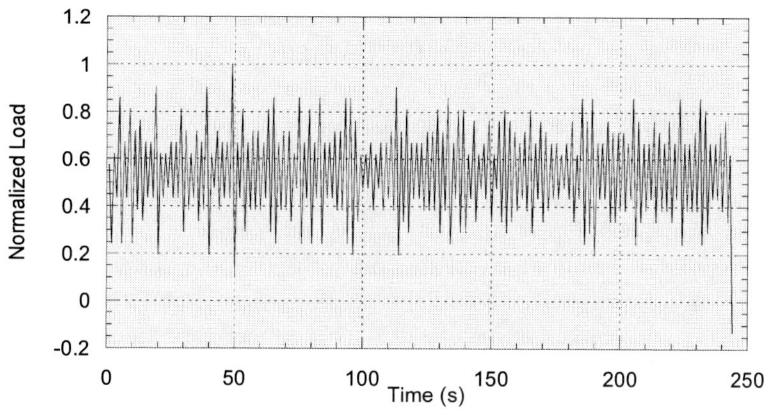

Figure 11. Typical flight loads.

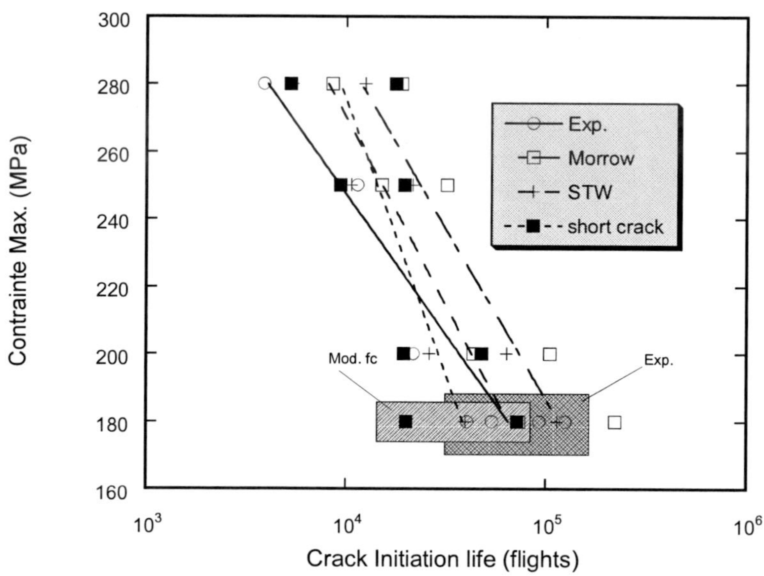

Figure 12. Flight simulation tests – comparison between experiments and predictions.

The test spectrum studies consisted of 1000 different flights and 22547 peaks. The next figure compares predicted lives using conventional models (using Miner's rule) and the short crack model. It is again confirmed that the short crack models prediction capacity is very good. Scatter in lives using this model is obtained by considering, different initial crack sizes.

DISCUSSION

The validity of the short crack model presented here has to be tested in different configurations and materials. We should look further into blunt notch configurations than the ones studied here. The blunt and sharp notch configurations could have different behaviors as reported in [12].

One of the assumptions in this model is that the short crack propagates without any closure effects and the crack at the notch behaves as the long crack behavior corrected for closure. This aspect was studied using marker band tests for cracks emanating from a notch in the 2024 alloy. The results are given in figure 13.

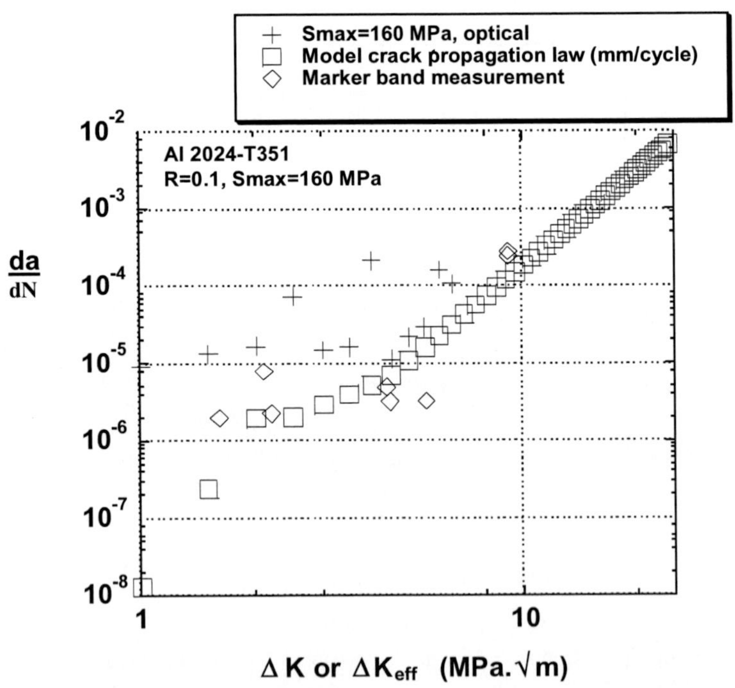

Figure 13. Comparison of crack behavior at the notch with long crack data.

It can be seen here that while optical measurements on the surface seem to indicate a short crack effect (crack growth rates higher than those of long cracks). The marker band spacings indicate that this effect is minimal and the cracks at the notch have behavior as predicted by the long crack data corrected for closure.

CONCLUSION

In this paper, a short review of conventional notch fatigue life evaluation is briefly presented. Marker band tests show that the crack initiation life at a notch can be neglected in some situation. A short crack model for predicting notch fatigue behavior has been tested in different configurations. The model predictions are quite acceptable as compared to measured fatigue lives.

REFERENCES

[1] Wanhill, R.J.H., *Milestone case histories in aircraft Structural integrity*, NLR-TP-2002-521, National Aerospace Laboratory, Netherlands.

[2] Pilkey, W.D., Petersons Stress Concentration Factors, John Wiley and Sons, NY, USA 1997.

[3] Peterson, R.E., Stress Concentration Design factors, John Wiley and Sons, NY, USA 1953.

[4] Glinka, G., Ott, W., and Nowack, H., "Elasto-Plastic Plane Strain Analysis of Stresses and Strains at the Notch Root" *Journal of Engineering Materials and Technology*, Vol. 110, 1988, N° 3, pp. 195-204.

[5] Wallbrink,C. and Hu, W., " A Strain Life Modula for CGAP, *Theory User Guide and Examples,* DSTO –TR- 2392, Air vehicles Division, Defense Science and Technology Organization, Australia, 2010.

[6] Newman, J.C. and Raju, I.S., "Stress-Intensity Factor Equations for Cracks in Three-Dimensional Finite Bodies Subjected to Tension and Bending Loads" *Computational Methods in the Mechanics of Fracture*, Atluri, Ed., Vol. 2 in *Computational Methods in Mechanics*, Elsevier Science Publ. Co., Inc., 1986, pp. 311-334.

[7] Kujawski, D., "Estimations of Stress Intensity Factors for Small Cracks at Notches" *Fatigue and Fracture Engineering of Materials and Structures*, Vol. 14, N° 10, 1991, pp. 953-965.

[8] Gérard, N., Ranganathan, N., Leroy, R., Mazari, M., and Bachir-Bouiadjira, B.A., "Fatigue Crack Initiation Life Estimation at a Notch: a New Software," *Applications of Automation Technology in Fatigue and Fracture Testing and Analysis: Fourth Volume, ASTM STP 1411,* A. A. Braun, P. C. McKeighan, A. M. Nicolson, and R. D. Lohr, Eds., American Society for Testing and Materials, West Conshohocken, PA, 2002.

[9] Aldroe, H., Tougui, A., Poirier, B., Lacroix, F., Venugopal, B., Mani, J.S., and Ranganathan, N. "Marker-bands techniques in fatigue", *Proceedings, DVM/ASTM Conference on Material and Component Performance on Variable Amplitude Loading*, Editors C.M.Sonsino and P.C. Mc Keighan, 2009 2, pp 967 – 976.

[10] Gérard, N., Leroy, R. , Girard, O., Ranganathan, N., "Crack Initiation at a notch under Constant and Selected Variable Amplitude Loading Conditions", Fatigue Testing and Analysis Under Variable Amplitude Loading conditions, *Journal of ASTM International*, vol.1, Issue 9, 2004 (Journal on line).

[11] Bleuzen, C., Chaudonneret, M., Farcy, L., Flavenot, J.F. and Ranganathan, N."Fatigue Testing and Life Prediction for Specimens of 2024 and 7010 alloys Subjected to Aeronautical Spectra, ASTM STP 1231, Philadelphia, USA, 1994, pp. 504-526.

[12] Veerman, Y. and Limondin, N., "Fatigue notch factor and Short Crack Propagation", *Enigennering Fracture Mechanics*, 2008, 75, pp 1320-1335.

In: Recent Trends in Fatigue Design
Editor: Ricardo Branco
ISBN: 978-1-61668-410-5

Chapter 4

Physical Damage Tolerant Design Approach for Elasto-Viscoplastic Cast Alloys Undergoing Thermomechanical Fatigue Loading (Application for Automotive Al-Si Cast Alloys Components)

Elias Merhy*
PSA Peugeot Citroën B.P. VV1415, Vélizy Villacoublay Cedex, France

Abstract

For elasto-viscoplastic cast alloys undergoing thermomechanical fatigue loading, crack initiation lifetime can be a small fraction of the total lifetime since even high quality cast alloys contain porosity, oxides and other inclusions. Accordingly, damage tolerant design approach has to be used in order to take into account the crack growth period in the total lifetime of the structure. Therefore, thermomechanical fatigue crack propagation behaviour has to be finely studied and understood as to develop adapted physically based non-isothermal crack growth law.
This chapter presents the development of a new damage tolerant design approach applied for Al-Si engine cylinder heads cast alloys. The work first step is to conduct several finite element simulations on a cylinder

* Corresponding Author address. Email: elias.merhy@mpsa.com ; elias.merhy@hotmail.com

head 3D model, to verify the localisation of the structure critical zone, and to determine the histories of stress and temperature fields in that zone. Then, crack growth mechanisms are experimentally analysed under different thermomechanical fatigue loading conditions. Tests are operated on different specimen types at different load ratios (*R*), temperatures (*T*), load frequencies (*f*) and stress intensity factor values (*K*) in order to establish the contribution of each parameter governing the fatigue crack growth. This analysis has led to the conclusion that the plastic strain range and the extent of the plastic zone ahead of the crack tip can be used to describe crack growth mechanisms and to estimate the cracking rate. Consequently, a physically based crack propagation law is developed on the base of these two physical quantities. The law consists on the integration of the inelastic strain amplitude at the crack tip region using a computational asymptotic solution for an elasto-viscoplastic material. The law can be applied for complex thermomechanical loading with positive or negative load ratio.

Keywords: Thermomechanical fatigue, Crack growth law, Damage tolerant design, Lifetime prediction, elasto-viscoplastic materials, Al-Si cast alloy

INTRODUCTION

Operating temperature of automotive Al-Si casting alloy cylinder heads can exceed 230°C in some zones. However, when the engine is off, its temperature could reach negative values in winter. This severe repeated temperatures variation corresponding to the engine start-stop cycles generates significant thermal strain fields that vary considerably between the centre and the borders of the flame deck of the cylinder head. As the nose bridge between the inlet and the exhaust valves (critical zone of the cylinder head) is quite embedded, the total strain field in that zone is considered null. Consequently, the mechanical strain, balancing the thermal strain, induces a thermomechanical fatigue loading in that zone. The loading goes essentially from tension at low temperature to compression at high temperature in the longitudinal direction of the nose bridge and could cause the nucleation and propagation of a crack from the flame deck toward the cooling system as shown in Figure 1.

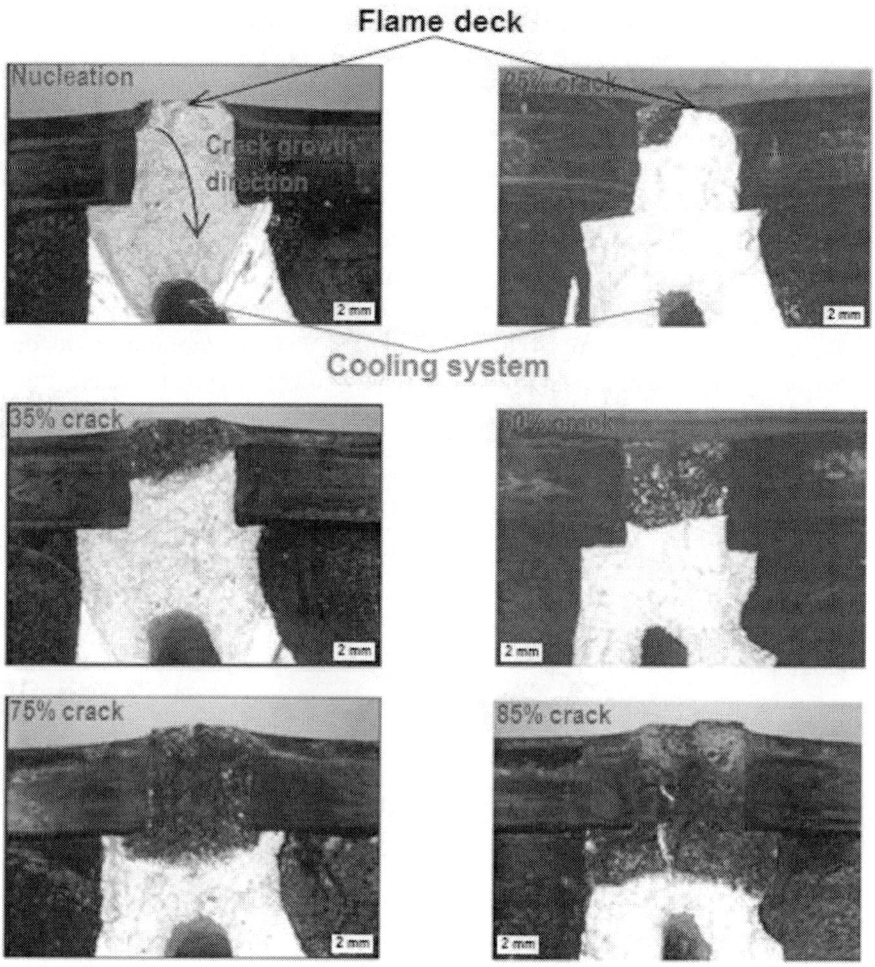

Figure 1. Crack nucleation and advance in an inlet/exhaust nose bridge of a HDI engine cylinder head subjected to an engine bench test.

Bench tests have been conducted on several engines. An engine bench test includes 10000 loading cycles, where for each cycle, the cylinder head reaches 230°C during 140s and then it is cooled rapidly within 80 seconds to reach 80°C. Then, the period of the loading cycle is equal to 220s which corresponds to a loading frequency of about 5.10^{-3} Hz. These tests have shown that the crack propagation lifetime is greater than that of the crack nucleation. Consequently, lifetime prediction law is optimal only if it is based on a "damage tolerant" approach considering both of initiation and propagation lifetime. However, cylinder heads are currently designed using the "safe life" approach (S-N curve) which only considers the crack nucleation lifetime and which is mostly based on the dissipated mechanical energy in the material $\left(\Delta W = \oint \sigma_{ij}(x,t) : \dot{\varepsilon}_{ij}(x,t)dt\right)$. Therefore, thermomechanical fatigue crack growth mechanisms and behaviour have to be perfectly understood in order to be able to develop and implement an adapted non-isothermal crack growth law within the framework of a damage tolerant design approach.

Operating at high temperature environment and having elasto-viscoplastic behaviour, Al-Si cylinder heads must be designed by considering the crack advance due not only to cyclic loading but also to monotonic loading. Accordingly, the new crack propagation law must figure all of the purely fatigue effect, the crack growth under monotonic loading condition (time dependent parameter) and the effect of the compressive load on the global crack advance, and all that, under non-isothermal condition.

Within this framework, several physical phenomena and parameters are involved in the crack propagation mechanisms, namely the crack closure phenomenon, the material microstructural features, loading frequency and temperature, compressive load and finally history effect. Here is a brief bibliographic review on the effect of each of these parameters on crack growth as it has been reported by several authors.

Crack Closure Phenomenon

Elber [1] was the first to introduce the crack closure concept. After that, many authors have focused their attention on the effective stress intensity factor range ΔK_{eff} calculation. Several crack closure mechanisms have been distinguished. Among them, the roughness induced crack closure has been attributed to the crack path deflections, especially at near threshold crack growth rates [2, 3]. Nonetheless, the plasticity induced crack closure is present over a large range of stress intensity factors and especially at high crack growth rates [1, 4]. The traditional concept of closure has led to the assumption that the driving force for crack propagation, ΔK_{eff}, exists only when the crack is fully open. However, more recent works [5, 6] have shown that the exclusion of the additional driving force below the crack opening load threshold can lead to inaccurate values of ΔK_{eff}. Therefore, partial crack closure models [5-8] and the two parameters driving force models [7, 9-12] have been developed.

Material Microstructural Features

In general the fatigue behaviour of aluminium casting alloys is dominated by the presence of casting defects that shortens not only the fatigue crack propagation period, but also the initiation period [13-17]. However, with improvement of casting quality and reduction in defect size (as in our case), the influence of microstructural features on fatigue performance cannot be neglected. Several works have shown that the fatigue crack growth behaviour of Al-Si alloys free of casting defects is governed essentially by the secondary dendrite arm spacing (SDAS) and by the size and shape of second phase such as eutectic Si particles [18-20]. The cracking and debonding of irregular shaped silicon particles is found likely to control the damage mechanisms of this biphasic alloy [21, 22]. This phenomenon has been observed in our study as well as a crack growth mechanism transition that has been associated to microstructural features influences.

Loading Frequency and Temperature

All these previous works treated essentially fracture mechanics problems under isothermal condition and without taking into account the time dependent crack growth. Few papers treating the effects of sustained load or loading frequency on cast aluminum alloys can be found in literature [23, 24]. However, since aluminium alloys are viscous materials, the loading frequency at high temperature could have an important effect on the crack growth rate. Our tests results have shown an important variation of the crack growth rate according to loading frequency, especially at high temperatures.

Compressive Load

The vast majority of fatigue crack growth laws found in literature assume that the crack advances only under remote traction loading [1, 5-12]. Hence, under negative load ratios, ΔK is considered equal to $K_{\max}$ assuming that the crack is closed and then does not advance. Although, over these last years, several authors have shown that the compressive loading accelerates fatigue crack growth [18-20, 25-27]. Furthermore, it has been reported that even under purely compressive loading, cracks are able to initiate and to grow until a given length [28-32]. This phenomenon has been explained by the existence of a local tensile residual stress field in the crack tip vicinity under remote negative loading. This same phenomenon has been observed in our tension-compression fatigue cracking tests and is found to be the reason of a crack growth rate increase compared to tests performed at the same level of K_{max} but at a positive load ratio ($R = 0.1$).

HISTORY EFFECT

Linear elastic fracture mechanics characterises the fracture behaviour of the materials that respond elastically under loading as for brittle and high strength materials. However, for low strength and high ductility alloys, the material shows a pronounced nonlinear behaviour in the crack tip plastic zone that generates the so called "history effect". In this case, linear elastic fracture mechanics characterisation of stress field in the crack tip vicinity may be faulty especially under a large number of cycles where errors are accumulated one cycle after another [21, 22, 25]. Then, a nonlinear ductile fracture mechanics approach must be used to better describe the fracture behaviour under elastic-plastic conditions, accounting then to the plasticity and the history effect.

METHODS

1. Structure Numerical Analyses

Three dimensional finite element simulations have been conducted on an engine bloc model consisting of a cylinder head and a cylinder bloc, using Abaqus software, cf. Figure 2a. The model consists of almost 400,000 ten-node tetrahedral *elements* and contains numerous surface contacts like those between the valve seats or the valve guides and the bloc of the cylinder head, and the contact between the crack free surfaces. Calculations have been conducted using a unified elasto-viscoplastic constitutive model with nonlinear kinematic hardening [33]. The model volume, the contact equations and the complexity of the material constitutive model make the numerical simulation heavier and lead to prohibitive computational time. For this reason, submodelling technique has been used in some cases for the inlet/exhaust nose bridge which is the structure critical zone.

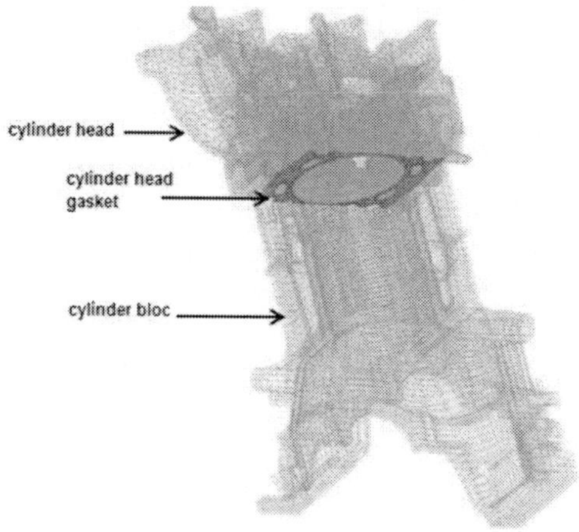

(a)

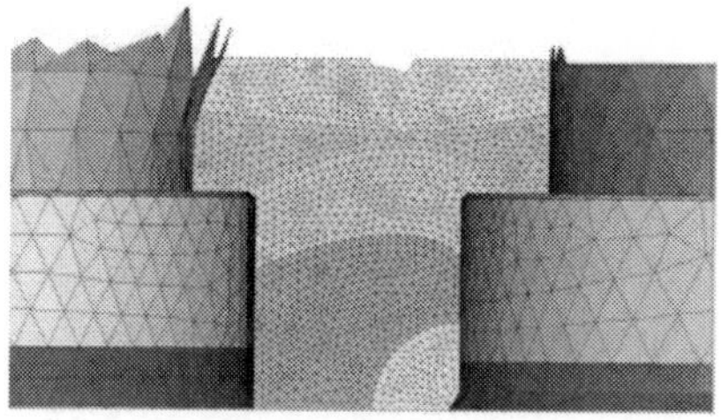

(b)

Figure 2. a) Finite element mesh of the 3D engine bloc model. b) Refined mesh in one of the nose bridge showing the different crack levels and shapes.

With the aim of obtaining accurate results of the mechanical fields in the structure critical zone, a mesh refinement of one of the nose bridges is realised with an element size almost equal to 200 μm, as shown in Figure 2b. In addition, different crack levels at 25, 50, 75 and 85% of the crack total length are created (cf. Figure 2b) and chosen in a way to correlate the crack growth shape observed experimentally and shown in Figure 1.

The loading cycles applied to the model are similar to that of the engine bench tests (140s of heating and 80s of cooling; $f \approx 0.005$ Hz). Numerical analyses of temperature, cumulative viscoplastic strain and total dissipated energy have been conducted on the cylinder head flame deck experiencing the higher levels of temperature. The cylinder head temperature reaches its maximum value in the centre of the flame deck, as shown in Figure 3a.

The temperature gradient from the centre to the borders of the flame deck causes the compressive stresses in the inlet/exhaust nose bridge at high temperature (engine on). While at low temperature (engine off), the valve seat radial stresses cause the tensile stress field on the nose bridge. The calculation of the accumulated plastic strain over a cycle $\left(\oint \sqrt{(2/3)\, \varepsilon_{ij}\!:\varepsilon_{ij}\, dt}\right)$ allows the detection of the critical zone that corresponds to the most plastically deformed area. Figure 3b shows clearly that the accumulated viscoplastic strain is essentially concentrated at the inlet/exhaust nose bridges, which reveals the criticality of these zones. Likewise, Figure 3c illustrating the dissipated energy $W\left(W = \oint \sigma_{ij}\!:\dot{\varepsilon}_{ij}\, dt\right)$ in the cylinder head flame deck, shows that this quantity, that characterises the material damage, reaches its maximum value in the inlet/exhaust nose bridges.

(a)

(b)

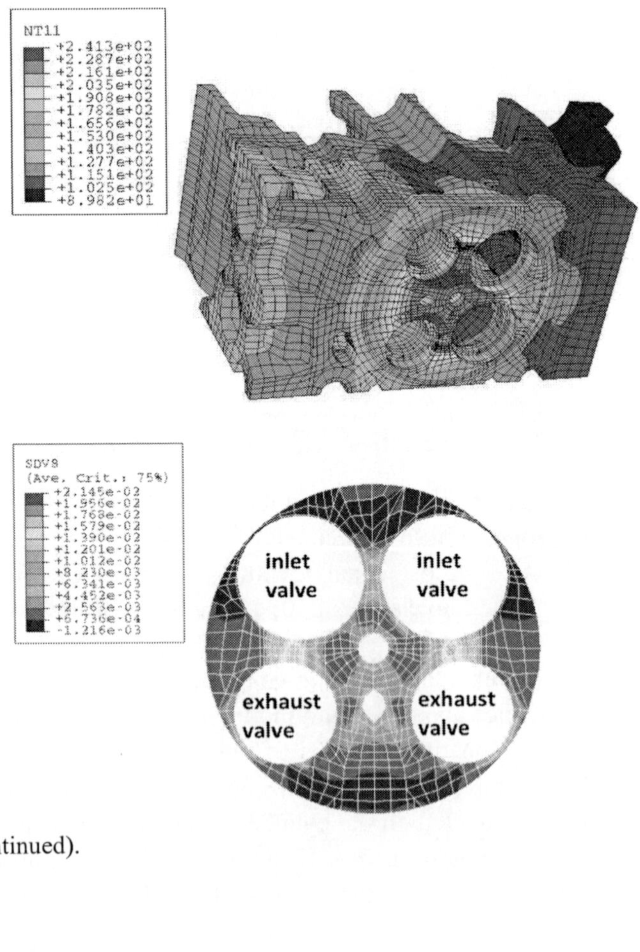

Figure 3. (Continued).

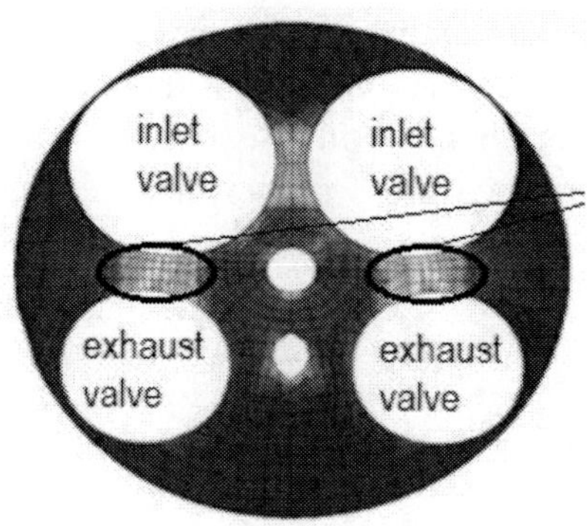

(c)

Figure 3. a) Temperature field in the cylinder head model obtained at the peak temperature of the cycle. b) Cumulative viscoplastic deformation field calculated over a complete cycle at the flame deck part of the cylinder head. c) Total energy dissipation field calculated over a complete cycle at the flame deck part of the cylinder head.

After verifying the criticality of the nose bridge zone, several local numerical analyses have been led in this zone in order to reveal the local cyclic stress-strain response of the material. Thus, stress, strain and temperature history have been calculated on fourteen solid elements chosen in the refined mesh nose bridge in a way to correlate the real crack advance trajectory, as shown in Figure 4a. It is worth emphasising that the stress and strain tensors have been calculated in local axis systems so as to obtain the tensors components in the exact opening (u axis), advancing (j or z axis) and shearing (k or v axis) directions.

The maximum temperature experienced by the lowest part of the nose bridge (E1: element on direct contact with the flames) is almost 230°C. While the peak temperature at the highest part (E14: element in direct contact with the cooling liquid) does not exceed 170°C, as shown in Figure 4b. Figure 5 shows the evolutions of σ_{uu} stress versus $\varepsilon_{\mathrm{uu}}^{\mathrm{vp}}$ inelastic deformation in the crack opening direction (u) for different solid elements of the nose bridge in a non-cracked cylinder head configuration. These results reveal a negative stress ratio in the critical zone ($R \approx$ -0.57). In fact, the stress in the (u) direction varies, going from traction at low temperature (first half of the cycle) to compression at high temperature (second half of the cycle). The material in the zone going from E1 to E9 presents an important cyclic viscoplastic response and a tendency to a plastic shakedown after 13 cycles (cf. Figure 5). The part going from E10 to E14 shows no more cyclic plastic response, the material in this zone presents an elastic thermomechanical behaviour manifested by a rapid elastic shakedown. In fact, this zone is the most cooled part of the bridge since E14 is in a direct contact with the cooling system. Then, it is slightly affected by the compressive stress due to thermal gradient. In another hand, this part is almost out of the tension stress influence induced by the valve seats at low temperatures.

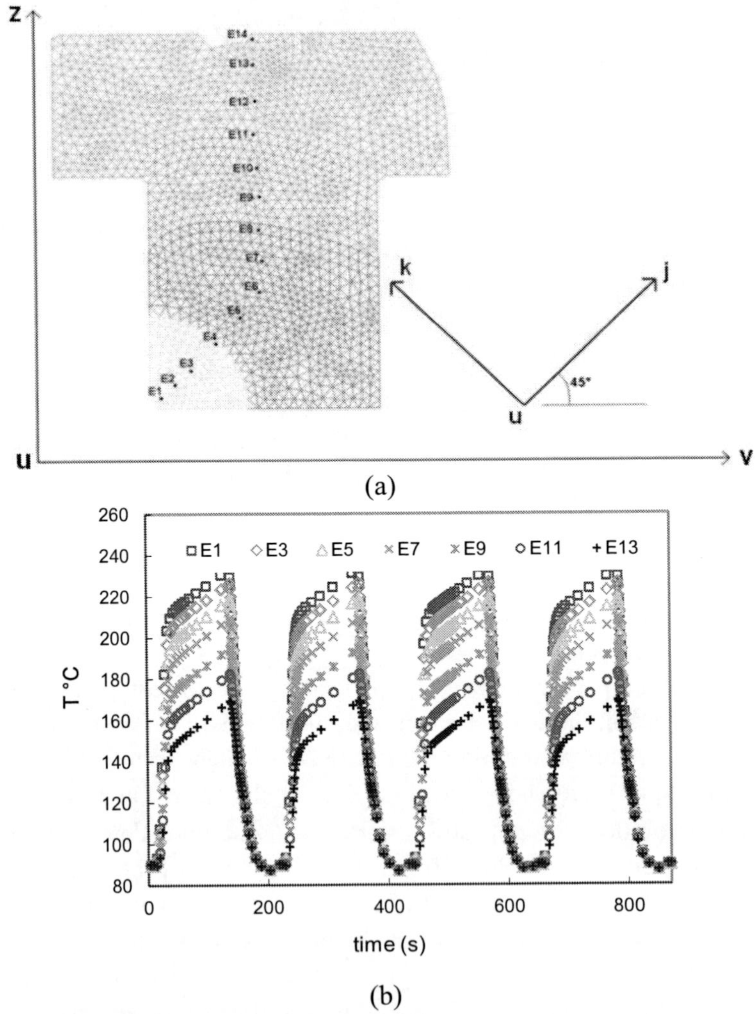

Figure 4. a) Solid elements in the nose bridge referenced to their local axis systems: E1 to E5 axis (u,j,k) ; E6 to E14 axis (u,v,z). b) Temperature evolution for 4 cycles in different solid elements of the nose bridge.

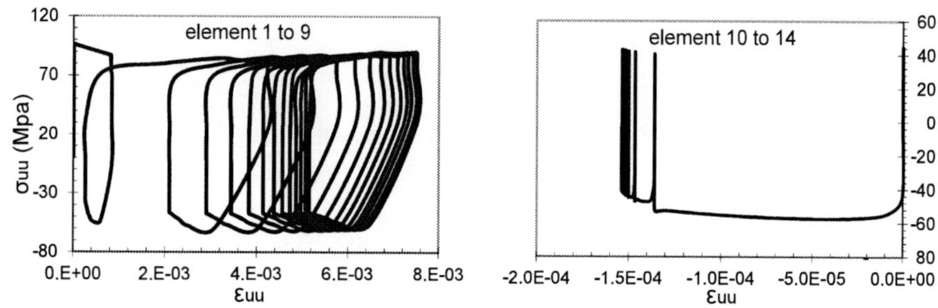

Figure 5. Evolution of σ_{uu} stress versus ε_{uu}^{vp} inelastic deformation for 13 cycles in different solid elements of the nose bridge.

Accordingly, the cylinder head nose bridge can be divided into two different parts having different thermomechanical behaviours:

- E1 to E9: dissipative zone characterised by generalised plasticity where the material shows a viscoplastic cyclic behaviour. In this zone the crack advances very rapidly. This makes unnecessary the analysis of this zone, since the quasi totality of the crack growth lifetime is spent in the other part of the nose bridge.

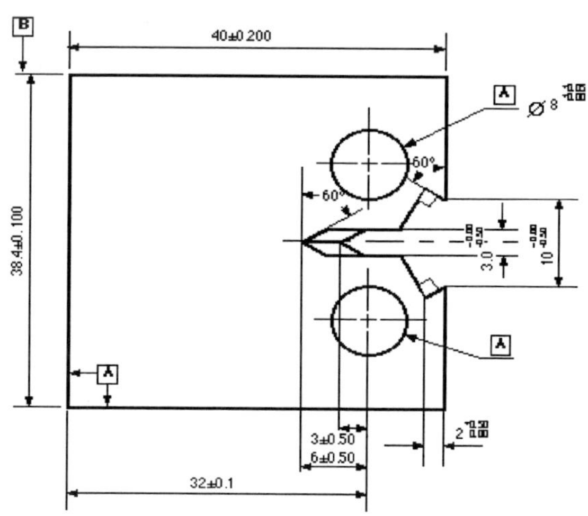

(a)

Figure 6. (Continued).

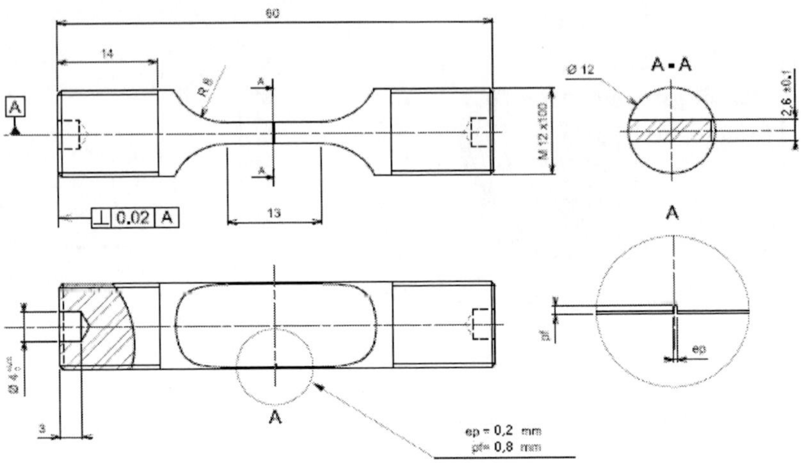

(b)

Figure 6. a) Plan of CT16 specimens used for fatigue cracking tests at $R > 0$. b) Plan of SEN specimens used for fatigue cracking tests at $R < 0$.

- E10 to E14: non-dissipative zone showing a confined plasticity thermomechanical non-isothermal response. For this region, characterised by a low crack growth rate, a fine analysis of crack growth mechanisms and behaviour is presented in the next section.

Using the results of these finite element calculations, the loading range and type in the cylinder head nose bridge is identified. Thus, suitable bench and specimen tests can be designed in order to better represent the real structure loading.

2. Crack Growth Experimental Analyses

Our experimental work is focused on studying long crack growth in the framework of confined plasticity which corresponds to the zone E10-E14 of the cylinder head nose bridge. A large number of fatigue cracking tests have been conducted on two types of specimens, compact tension (CT16) and single edge notch (SEN) shown in Figure 6a and Figure 6b respectively. For the CT16 specimens, several positive load ratio cracking tests have been conducted, some under increasing ΔK and different load ratios, temperatures and frequencies and others under constant ΔK and variable frequencies, load ratios and temperatures. In total, these tests cover forty one loading conditions.

The SEN specimens have been used for negative load ratio tests. Using this specimen type, numerous constant ΔK tests have been conducted at different temperatures, ΔK, load ratios and frequencies. In addition, two non-isothermal increasing ΔK tests, representing the real loading conditions experienced by the nose bridge, have been carried on the SEN specimens. In total, the SEN tests characterise fifty loading conditions. A summary of all conducted tests is presented in the table below.

The full experimental investigation of crack growth mechanisms of Al-Si alloys is reported in a previous work [34]. Here, the most significant outcomes of this work are recalled.

Scanning electron microscopy observations of the specimens crack surfaces of the tests conducted at positive load ratios have revealed a mixed mechanism crack growth, where fatigue traces and microvoids (i.e. ductile fracture) are present at the same time. It has been shown that cracking growth has a progressive mixed mechanism that goes from a total fatigue crack advance at low stress intensity factor values (beginning of the Paris regime) to a mixed fatigue/static load crack growth at the middle of the Paris regime, and finally to a total crack propagation under static load at the end of the Paris regime. In addition, it has been shown that fatigue cracking is essentially governed by the K_{max} and the material microstructural features (SDAS, and Si particles shape and size). Indeed, when the crack tip plastic zone size r_p is smaller than the SDAS (low K_{max} value) the plastic strain energy is not sufficient to damage particles and the crack has an almost plane path. However, when r_p is larger than the SDAS, the plastic strain energy become sufficient to damage particles (by cracking or debonding the particles from the Al matrix). Then, the crack follows the damaged inter-dendrite eutectic zones, corresponding to cracked or debonded particles, making a tortuous but more energetically favorable path. Thus, this crack mechanism and path changing phenomenon has been explained by the crack tip material damage at the Al/Si interfaces. The damage amount has been characterised by the amplitude and the extent of the crack tip plasticity.

Fatigue cracking tests have shown that the crack propagation speed of the Al-Si alloy is largely dependent on the loading frequency *f*. Generally, crack growth rate increases when the frequency decreases. The parameters governing the crack propagation under positive load ratio are K_{max} and T at low frequency and ΔK and T at high frequency.

For high stress intensity factor values, important crack tip plasticity leads to the loss of the stress singularity of order $r^{-1/2}$ and introduces a history effect i.e. at a given cycle, cracking mechanism and rate become depending on plasticity induced by previous cycles. Accordingly, crack tip cumulative plasticity (history effect) can cause important crack growth rate increase.

Table 1. Summary of the tests conducted on both specimens types CT16 and SEN

test no.	R	specimen	ΔK (MPa$\sqrt{m}$)	T (°C)	f (Hz)
1	0.1	CT	increasing	20	20
2	0.1	CT	increasing	100	20
3	0.1	CT	increasing	150	20
4	0.1	CT	increasing	150	25
5	0.7	CT	increasing	100	20
6	0.1	CT	8	20	20-1-0.05-0.005
7	0.1	CT	8	100	20-1-0.05-0.005
8	0.1	CT	8	150	20-1-0.05-0.005
9	0.1	CT	12	20	20-1-0.05-0.005
10	0.1	CT	12	100	20-1-0.05-0.005
11	0.1	CT	12	150	20-1-0.05-0.005
12	0.7	CT	5.5	20	20-1-0.05-0.005
13	0.7	CT	5.5	100	20-1-0.05-0.005
14	0.7	CT	5.5	150	20-1-0.05-0.005
15	-0.5	SEN	13.5	20	20-1-0.05-0.005
16	-0.5	SEN	13.5	100	20-1-0.05-0.005
17	-0.5	SEN	13.5	150	20-1-0.05-0.005
18	-0.5	SEN	20	20	20-1-0.05-0.005
19	-0.5	SEN	20	100	20-1-0.05-0.005
20	-0.5	SEN	20	150	20-1-0.05-0.005
21	-1	SEN	18	20	20-1-0.05-0.005
22	-1	SEN	18	100	20-1-0.05-0.005
23	-1	SEN	18	150	20-1-0.05-0.005
24	-1	SEN	26.6	20	20-1-0.05-0.005
25	-1	SEN	26.6	100	20-1-0.05-0.005
26	-1	SEN	26.6	150	20-1-0.05-0.005
27	-0.57	SEN	increasing	$90<T<177$	0.005
28	-0.57	SEN	increasing	$90<T<177$	0.005

Negative load ratios tests have shown that compressive parts of the loading cycles cause important crack growth rate increase due to a local plasticity phenomenon at the crack tip eutectic particle region that induces local residual tension stress fields. It has been shown that the remote compressive load effect on the global crack growth is subjected to a saturation or limitation phenomenon. Indeed, the crack growth rate due to negative load is very high when the frequency is high and K_{max} is low, however, negative load crack growth becomes negligible at high temperature and K_{max} values and low frequency. Thus, it has been proved that, for this material and at negative load ratios, classical crack closure concepts (ΔK_{eff}) are inaccurate to predict crack propagation rate.

The major result of the experimental work is that fatigue crack growth mechanisms and rate are governed by the material microstructural features damage that can be characterised and estimated by using the crack tip plastic zone size and the inelastic deformations range in that zone [34]. Based on this final result, the new crack growth law has been developed.

3. Physically Based Crack Growth Law for Damage Tolerant Design Approach

In a highly competitive automotive environment, manufacturers work hardly to reduce the vehicles development cost and to offer the customers the best cost/benefits compromise while respecting safety and pollution standards increasingly stringent. These strategic objectives are achieved through a policy of vehicles mass reduction that aims to limit fuel consumption and pollutant emissions. Meanwhile, engines performance is constantly increasing. Consequently, engines components are increasingly loaded, particularly for cylinder heads and exhaust manifolds whose materials are pushed to their limits. Thus, these components should be optimally designed in order to meet the above requirements while remaining functional and fulfilling their desired lifetime. To do so, and in order to reduce design costs, scientific mastery of the materials damage mechanisms becomes primordial today, so as to be able to precisely predict the fatigue lifetime.

Within this framework, damage tolerant design methods are the best to be applied, since they are optimal approaches that estimate the structure lifetime by taking into account both of crack initiation and crack growth lifetime. As the safe life approach is already used and mastered, this work has been focused on the crack growth phase. In the context of elasto-viscoplastic materials undergoing negative load ratio non-isothermal loading, physically based crack growth laws are the most appropriate. Indeed, this type of model can be easily generalised to the case of complex loading because it is often based on a global energy approach considering intrinsically all types of solicitations. Thus, these laws are able to take into account the interaction between different damage mechanisms due to fatigue, oxidation or creep. Therefore, the main objective of this work is to develop a physically based crack growth law allowing the evaluation of the crack propagation lifetime for complex material behaviour and loading conditions and taking into account all the effects of cyclic loading, monotonous (time effect) and compressive load.

Fatigue cracking tests and experimental analysis have shown that the viscoplastic strain range and the inelastic zone size are the major physical quantities governing the crack propagation for the Al-Si alloys. Then, in order to develop a new crack propagation law based on these variables, it is necessary to characterise the local material behaviour at the crack tip using refined finite element analysis and then to numerically calculate these quantities. This can lead to develop analytical formulations of the viscoplastic strain range and the crack tip inelastic zone size.

3.1. Material Behaviour Analysis in the Crack Tip Vicinity

A refined finite element model (mesh size ≈ 10 μm) has been used for this analysis. Mechanical fields (stress tensor σ_{ij}, deviatoric stress tensor S_{ij}, hydrostatic stress P and von Mises equivalent inelastic deformation ε_{eq}^{vp}) has been calculated in the model median plane in the vicinity of the crack tip at different points defined by their polar coordinates r and θ (cf. Figure 7), and that for different loading conditions (T, R, K_{max}, and f). Graphics of Figure 7 represent some of those results. The material shows a plastic shakedown behaviour even at only 5μm ahead of the crack tip. The equivalent inelastic strain range ($\Delta\varepsilon_{eq}^{vp}$) has been stabilised after almost twenty cycles. Then, in what follows, $\Delta\varepsilon_{eq}^{vp}$ will denote the equivalent viscoplastic strain range [35] calculated over the 20th cycle as follows:

$$\Delta\varepsilon_{eq}^{vp} = \max_{t_1} \max_{t_2} \sqrt{\frac{2}{3}\left(\underline{\underline{\varepsilon}}_{t_1}^{vp} - \underline{\underline{\varepsilon}}_{t_2}^{vp}\right) : \left(\underline{\underline{\varepsilon}}_{t_1}^{vp} - \underline{\underline{\varepsilon}}_{t_2}^{vp}\right)} \tag{1}$$

Graphics of Figure 7 show a material behaviour changing according to both parameters r and θ. In fact, these curves show that, at constant r, $\Delta\varepsilon_{eq}^{vp}$ and P vary according to the angle θ, as illustrated in Figure 8.

Figure 8 shows that for different loading conditions, the equivalent inelastic deformation range increases with the augmentation of θ from 0° to 67° where it reaches its maximum value and then decreases with the increase of θ from 67° to 90°, while the hydrostatic stress reaches its maximum value when θ is equal to 0°. Figure 7 shows also an influence of the angle θ on the ratcheting observed before the plastic shakedown after a given number of cycles. In fact, when θ is lower than 67° a positive ratcheting deformation occurs, while a negative ratcheting takes place when θ is higher than 67°. Thus, the variation of the material thermo-mechanical response and the amplitude of inelastic deformations according to r and θ can be represented schematically as shown in Figure 9.

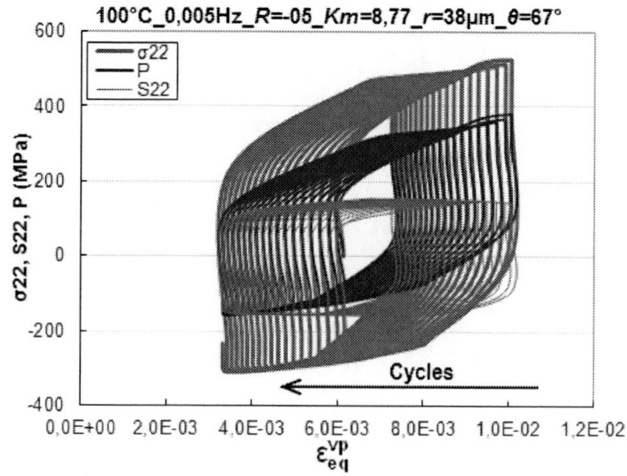

Figure 7. Evolution of σ_{22}, S_{22} and P according to ε_{eq}^{vp} on 19 cycles (2nd to 20th) for various points M (r, θ) and for different loading conditions.

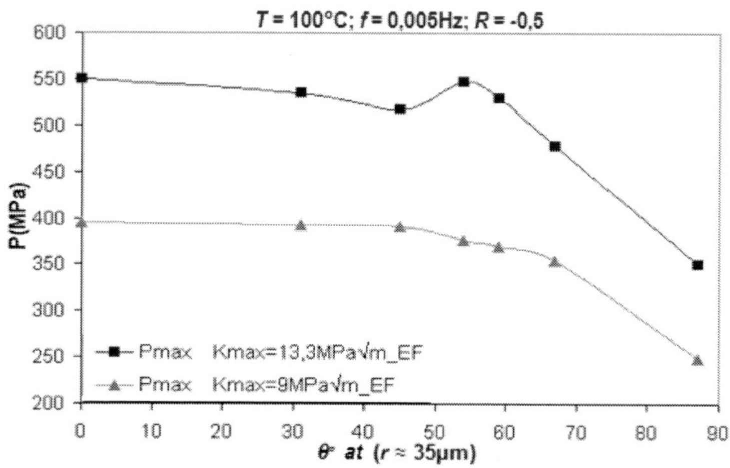

Figure 8. Evolution of $\Delta\varepsilon_{eq}^{vp}$ and the hydrostatic stress P (obtained by finite element calculations on the twentieth cycle) according to the angle θ for $r = 35$ µm.

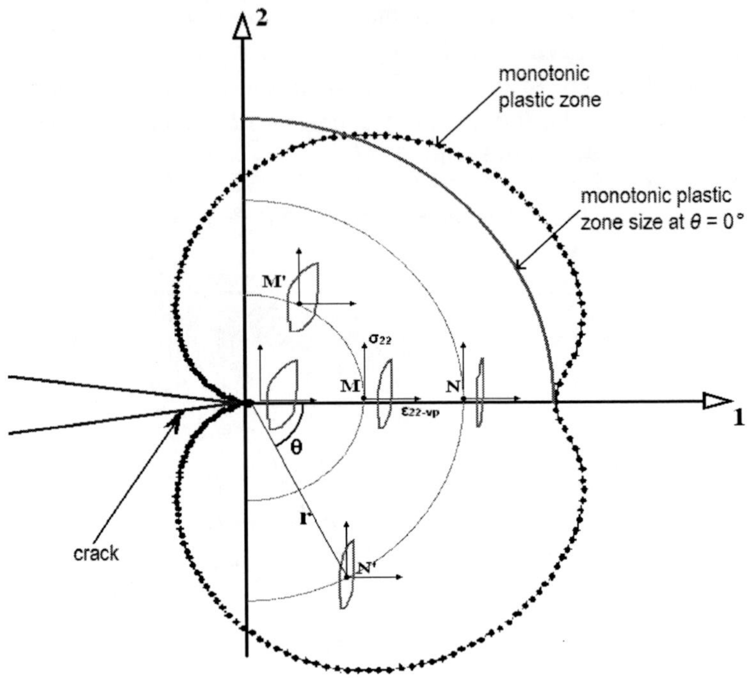

Figure 9. Schematic representation of the material behaviour and the variation of $\Delta\varepsilon_{eq}^{vp}$ at different points in the plastic zone (M, M' and N, N').

Figure 7 shows that, variations of σ_{22} and P are similar while S_{22} remains almost constant. This means that with the variation of the loading conditions, it is the hydrostatic stress that changes, and therefore this parameter governs the damage in the crack tip vicinity. This result confirms that the damage mechanism at the crack tip is characterised by decohesion between the particles and the matrix and therefore by a cavity growth depending essentially on P. This may explain the quasi-plane crack propagation (without macroscopic deviation) observed experimentally, since the hydrostatic stress P reaches its maximum value at $\theta = 0$ ° in front of the crack tip (cf. Figure 8). This type of damage suggests that the triaxiality ratio $\left(\chi = P/\sigma_{eq}\right)$ would have the major effect on material damage and then on the propagation rate. Hence, calculations of the triaxiality ratio are conducted for different loading conditions and partially presented in Figure 10.

Results of Figure 10 show that the triaxiality ratio varies slightly with either the loading conditions or r and θ. Thus by considering χ constant, the cavity growth damage, which is usually characterised for ductile materials by the Rice and Tracey law $\left(\dot{D} = 0.283\, d\varepsilon_{eq}\exp(\frac{3}{2}\chi)\right)$, becomes only dependent on the equivalent deformation $\left(d\varepsilon_{eq}\right)$ calculated in the extent of the plastic zone. This result can be considered as a validation of the approach adopted in this work for characterisation of thc crack growth mechanisms and rate.

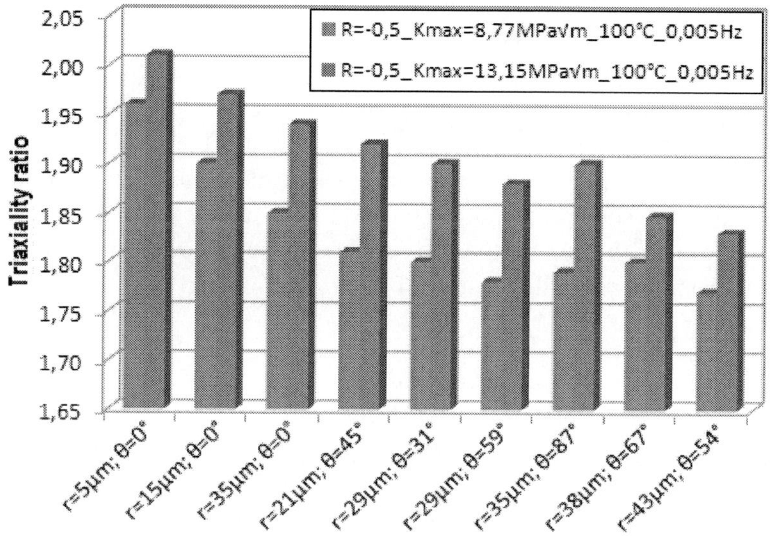

Figure 10. Evolution of the triaxiality ratio according to r, θ and loading conditions (results obtained at maximum load of the twentieth cycle).

3.2. Analytical Development of the Crack Growth Law

Crack growth rate prediction based on crack tip inelastic strain has been first introduced by McClintock [36]. Then, it has been the centre of interest of many authors that have conducted several crack growth analytical studies [37-41] or experimental ones [42-45]. Generally, these methods use an accumulation parameter based on the inelastic strain in a confined zone at the rack tip vicinity, where the elasto-plastic mechanical fields in the crack tip region are analytically or numerically calculated. The approach proposed by McClintock introduces a process zone of size ρ in which the accumulation parameter is considered as the mean value of the inelastic strain obtained by approximate solution of the deformation field in front of the crack tip. Within this framework and between all the methods proposed in the literature, the most suitable models are those based on a damage accumulation process zone whose size is adjusted to correlate the experimental results.

It should be remembered that, in our study case, cracks are driven by the effect of fatigue and monotonic loading at the same time. In addition, the material is submitted to complex negative load ratio non-isothermal loading conditions. Consequently, a physically based crack growth law has to be developed in order to intrinsically consider all these effects. Thus, based on the numerical and experimental analyses above, the new law is based on the summation of the equivalent inelastic strain range in front of the crack tip inside a large process zone. This zone corresponds to a semicircle having its centre at the crack tip and a radius equal to the monotonic plastic zone size in front of the crack tip at $\theta = 0°$, $r_{\mathrm{p}}^{\mathrm{mon}}(\theta = 0°)$ (cf. Figure 9). By considering the symmetry of the damaged zone in front of the crack tip according to the crack plane, the semicircle can be reduced to a quarter of a circle (cf. Figure 9). Then, the crack growth rate is calculated as being proportional to the total cyclic inelastic strain in the quarter of circle, as follows:

$$\frac{da}{dN} = \zeta \int_0^{\frac{\pi}{2}} \int_0^{r_{\mathrm{p}}^{\mathrm{mon}}(\theta=0)} \Delta\varepsilon_{\mathrm{eq}}^{\mathrm{vp}}(r,\theta) r dr d\theta \tag{2}$$

ζ is a proportionality factor which can be calibrated from a single isothermal fatigue cracking test. From a physical point of view, the proposed law considers that for any applied loading type, the total amount of viscoplastic strain corresponding to a given crack advance is independent of the "manner" in which the dissipation has been generated (whether by fatigue or monotonic loading).Therefore, the proposed law intrinsically takes into account the two effects of fatigue and static loading at positive or negative load ratios.

If $\Delta\varepsilon_{\mathrm{eq}}^{\mathrm{vp}}(r,\theta)$ and $r_{\mathrm{p}}^{\mathrm{mon}}(\theta = 0°)$ are directly calculated by finite element method, Eq. 2 can directly be resolved. However, these parameters cannot be directly obtained by finite element method except in the case of a very refined mesh model (mesh size ≈ 10 µm) in the vicinity of the crack tip. Applying such a refined mesh in the cracked regions of an engine finite element model including contacts, variable loading conditions and elasto-viscoplastic behaviour is obviously a complex task that leads to prohibitive computation times or even to calculation divergence. Then, in order to avoid the very refined mesh in real structure models, analytical solutions of $\Delta\varepsilon_{\mathrm{eq}}^{\mathrm{vp}}(r,\theta)$ and $r_{\mathrm{p}}^{\mathrm{mon}}(\theta = 0°)$ are developed in next sections using loading parameters (like *K*, *T*, *f*,...) that may be obtained from a relatively coarse mesh. This analytical approach is based on a computational asymptotic solution of $\varepsilon_{\mathrm{eq}}^{\mathrm{vp}}(r,\theta)$ under elasto-viscoplastic behaviour from which $\Delta\varepsilon_{eq}^{vp}(r,\theta)$ and $r_p^{mon}(\theta = 0°)$ can be estimated.

3.2.1. Asymptotic Solution *for* Equivalent Cyclic Viscoplastic Strain Field

Crack tip strain singularity for materials having nonlinear elastoplastic behaviour was first introduced by Hutchinson [46], Rice and Rosengren [47] (HRR fields). Hoff [48] has shown that by replacing the deformation for elastic-plastic material by the strain rate for viscoelastic material in all the basic constitutive equations, the asymptotic solution of the mechanical fields becomes similar for both types of materials. Thus, using the analogy of Hoff, Riedel and Rice [49] have developed an asymptotic solution, for a viscoelastic material with a Norton power creep, whose one-dimensional behaviour is expressed as:

$$\dot{\varepsilon} = \frac{\dot{\sigma}}{E} + \dot{\varepsilon}_0 \left(\frac{\sigma}{\sigma_0} \right)^n \tag{3}$$

E is the Young modulus, σ_0 and $\dot{\varepsilon}_0$ are respectively reference stress and strain rate and *n* is the exponent of the power creep law. The strain rate field viscoelastic asymptotic solution in the crack tip region developed by Riedel and Rice [49] is given by:

$$\dot{\varepsilon}_{ij} = \dot{\varepsilon}_0 \left(\frac{C(t)}{\dot{\varepsilon}_0 \sigma_0 I_n r} \right)^{\frac{n}{n+1}} \tilde{\varepsilon}_{ij}(\theta, n) \tag{4}$$

where I_n and $\tilde{\varepsilon}_{ij}$ are dimensionless functions of *n* and (*θ, n*) respectively. These functions are tabulated by Shih [50] or also by Chao and Zhang [51]. *C*(*t*) is a factor depending on the creep time, the amplitude of the macroscopic loading, the configuration of the crack and the material properties. It is defined by Bassani and McClintock [52] by the following contour integral:

$$C(t) = \int_{\Gamma} \left(\frac{n}{n+1} \sigma_{ij} \dot{\varepsilon}_{ij} n_x - \sigma_{ij} n_j \frac{\partial \dot{u}_i}{\partial x} \right) ds \tag{5}$$

where Γ is an open and oriented contour having its extremities situated on the top and the bottom crack flanks, x is the crack growth direction axis (direction 1 in Figure 9), u is the displacement vector at a given point of Γ and n_j is the jth component of the external unit normal vector to the outline contour. In the case of short term creep (primary creep), elastic strain is significantly greater than the strain due to creep with the exception of a very small area just in front of the crack tip and named "creep zone". Considering that elastic fields remain dominant outside the small primary creep area, Rice [47] has developed the following simplified expression of $C(t)$:

$$C(t) = \frac{K_I^2(1-\nu^2)}{E(n+1)t} \tag{6}$$

where ν is the Poisson ratio. Furthermore, several authors have studied the crack tip equivalent strain range field $\left(\Delta\varepsilon_{\text{eq}}(r,\theta)\right)$ and have proposed asymptotic solutions under elasto-plastic behaviour [53-55]. Rice [53] has considered that, under cyclic loading, $\Delta\varepsilon_{\text{eq}}(r,\theta)$ singularity can be deduced from that of ε_{ij} by replacing K with ΔK and σ_0 by $\Delta\sigma_0$:

$$\Delta\varepsilon_{\text{eq}}(r,\theta) = \frac{\Delta\sigma_0}{E}\left(f_n(\theta)\frac{\Delta K^2}{\Delta\sigma_0^2}\frac{1}{r}\right)^{\frac{n}{n+1}} \tag{7}$$

$f_n(\theta)$ is a dimensionless angular function of n and θ. All asymptotic stress and strain fields solutions found in literature are suitable for materials with elasto-plastic or viscoelastic behaviour.

In the present work, an asymptotic solution of $\Delta\varepsilon_{\text{eq}}^{\text{vp}}(r,\theta)$ based on the material constitutive model equations is developed for elasto-viscoplastic materials.

As a *first step*, the problem is treated by considering $\theta = 0°$. The material viscoplastic strain rate expression is given by:

$$\dot{\varepsilon}_{\text{eq}}^{\text{vp}} = \sqrt{\frac{2}{3}\dot{\underline{\underline{\varepsilon}}}^{\text{vp}} : \dot{\underline{\underline{\varepsilon}}}^{\text{vp}}} = \left\langle \frac{J_2\left(dev\left(\underline{\underline{\sigma}} - \underline{\underline{X}}\right)\right) - \sigma_y(T)}{\eta(T)} \right\rangle^{m(T)} \tag{8}$$

where $\underline{\underline{X}}$ is the kinematic hardening stress tensor, σ_y the yield stress, $\eta(T)$ and $m(T)$ are the viscosity parameters dependent on temperature. J_2 is the second invariant of the deviator of the tensor ($\underline{\underline{\sigma}} - \underline{\underline{X}}$) given by:

$$J_2\left(dev\left(\underline{\underline{\sigma}} - \underline{\underline{X}}\right)\right) = \sqrt{\frac{3}{2} dev\left(\underline{\underline{\sigma}} - \underline{\underline{X}}\right) : dev\left(\underline{\underline{\sigma}} - \underline{\underline{X}}\right)} \tag{9}$$

In the case of uniaxial non-isothermal loading (observed in the cylinder head nose bridge), J_2 is equal to $(\sigma - X)$. Thus, Eq. 8 can be expressed as follows:

$$\dot{\varepsilon}_{\text{eq}}^{\text{vp}} = \left\langle \frac{\sigma - X - \sigma_y\left(T(t)\right)}{\eta\left(T(t)\right)} \right\rangle^{m(T(t))} = \left\langle \frac{\sigma_{\text{v}}}{\eta\left(T(t)\right)} \right\rangle^{m(T(t))} \tag{10}$$

where σ_{v} is the viscous stress. Eq. 10 presents a power creep law identical to the viscous part of Eq. 3 used by Riedel and Rice. Thus, by analogy to Eq. 4 and Eq. 6 applied for primary creep stage, an asymptotic solution of the equivalent viscoplastic strain rate for non-isothermal condition $T(t)$ can be expressed for $\theta = 0°$ by:

$$\dot{\varepsilon}_{\text{eq}}^{\text{vp}}(r,t) = \dot{\varepsilon}_0 \left(\frac{K_{\text{Iv}}^2(t)\left(1-\nu^2\right)}{\dot{\varepsilon}_0 \sigma_y\left(T(t)\right).r.E\left(m\left(T(t)\right)+1\right).t.\eta\left(T(t)\right)} \right)^{\frac{m(T(t))}{m(T(t))+1}} \tag{11}$$

$\dot{\varepsilon}_0$ is a reference strain rate, and K_{Iv} is the viscous stress intensity factor given by:

$$K_{\text{Iv}} = \sigma_{\text{v}}(r,t,T) f(a) = \sigma(r,t,T) f(a) \frac{\sigma_{\text{v}}(r,t,T)}{\sigma(r,t,T)} = K_{\text{I}} \frac{\sigma_{\text{v}}(r,t,T)}{\sigma(r,t,T)} \tag{12}$$

$f(a)$ is a function characterising the structure geometry surrounding a crack of length a. When Eq. 12 right term replaces K_{Iv} in Eq. 11, we can notice that this last equation represents a nonlinear differential equation in ε_{eq}^{vp} and t. This nonlinear relationship cannot be solved analytically. Yet, the ratio $\frac{\sigma_v}{\sigma}$ can be expressed as a function of time (t) by using numerical calculations based on the constitutive model equations:

$$\frac{\sigma_v}{\sigma}(t) = 0.85\frac{t - t_{dp}}{t_m - t_{dp}} \quad ; t \in [t_{dp}, t_m] \tag{13}$$

where t_m is the instant of the stabilised loading cycle corresponding to the maximum loading and t_{dp} is the instant corresponding to the beginning of the inelastic strain increase in the ascending phase of the loading cycle (cf. Figure 11). By substituting K_{Iv} in Eq. 11 using Eq. 12 and Eq. 13, analytical solution of the differential equation becomes possible. Then, the equivalent viscoplastic strain can be deduced at any instant t of the cycle and that by integrating Eq. 11 in the interval of time going from t_{dp} to t, as follows:

$$\varepsilon_{eq}^{vp}(r,t) = \int_{t_{dp}}^{t} \dot{\varepsilon}_0 \left(\frac{K_I^2(t)\left(0.85\frac{t - t_{dp}}{t_m - t_{dp}}\right)^2 (1-\nu^2)}{\dot{\varepsilon}_0 \sigma_y(T(t)).r.E(m(T(t))+1).t.\eta(T(t))} \right)^{\frac{m(T(t))}{m(T(t))+1}} dt \ ;$$

$$t \in [t_{dp}, t_m] \tag{14}$$

Thus, the equivalent viscoplastic strain range can be defined as following:

$$\Delta\varepsilon_{eq}^{vp}(r) = \max_{t_1}\max_{t_2}\left(\varepsilon_{eq}^{vp}(r,t_1) - \varepsilon_{eq}^{vp}(r,t_2)\right) \tag{15}$$

Knowing that $\varepsilon_{eq}^{vp}(r,t)$ is a continuous and increasing function over $[t_{dp}, t_m]$, $\Delta\varepsilon_{eq}^{vp}(r)$ is expressed as:

$$\Delta\varepsilon_{eq}^{vp}(r) = \varepsilon_{eq}^{vp}(r,t_m) - \varepsilon_{eq}^{vp}(r,t_{dp}) = \varepsilon_{eq}^{vp}(r,t_m) \tag{16}$$

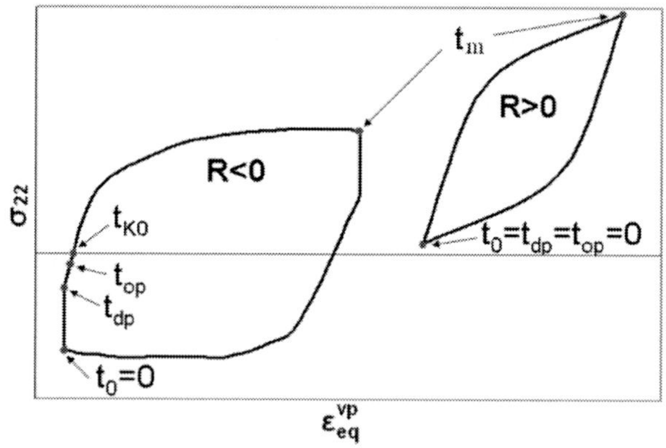

Figure 11. Schematic representation illustrating t_{dp}, t_m, t_{op} (instant of crack opening), and t_{K0} (instant for which K is equal to 0) for two cases of negative and positive load ratio cycles.

Hence one gets at $\theta = 0°$:

$$\Delta\varepsilon_{eq}^{vp}(r,\theta{=}0°) = \int_{t_{dp}}^{t_m} \dot{\varepsilon}_0 \left(\frac{K_I^2(t)\left(0.85\dfrac{t-t_{dp}}{t_m-t_{dp}}\right)^2 (1-\nu^2)}{\dot{\varepsilon}_0 \sigma_y(T(t)).r.E(m(T(t))+1).t.\eta(T(t))} \right)^{\frac{m(T(t))}{m(T(t))+1}} dt \quad (17)$$

With this result, the asymptotic solution of the cyclic equivalent inelastic strain for elasto-viscoplastic material is obtained for $\theta = 0°$. This solution is restrictively based on the material constitutive model equations. Thus, for any elasto-viscoplastic material whose constitutive model is known, $\Delta\varepsilon_{eq}^{vp}(r)$ can be calculated numerically, without any need to refer to new experimental data.

As a *second step*, the effect of the angle θ is introduced and a global asymptotic solution of $\Delta\varepsilon_{eq}^{vp}(r,\theta)$ for elasto-viscoplastic material is developed. Based on numerical analyses, a new model of inelastic strain rate evolution according to the angle θ has been developed for the crack tip zone. In fact, finite element simulations have shown that the variation of $\ln(\Delta\varepsilon_{eq}^{vp})$ according to θ follows systematically a second degree polynomial function that varies according to loading parameters by multiplying the hall polynomial equation by a factor λ dependent on these parameters. Accordingly, $\Delta\varepsilon_{eq}^{vp}$ can be written as a function of r, θ and the loading parameters as follows:

$$\ln\left(\Delta\varepsilon_{eq}^{vp}(r,\theta)\right)=\left[a\theta^2+b\theta+c\right].\lambda\left(K,T,f,R\right) \tag{18}$$

a, b and c are three numerically calibrated parameters. By applying this last equation for the particular case $\theta = 0°$, the parameter λ representing the loading condition effect can be identified as following:

$$\lambda(K,T,f,R)=\frac{1}{c}\ln\left(\Delta\varepsilon_{eq}^{vp}\left(r,\theta=0°\right)\right) \tag{19}$$

where $\Delta\varepsilon_{eq}^{vp}(r,\theta = 0°)$ is obtained from Eq. (17). Finally, the general form of the equivalent cyclic inelastic strain at any point in the crack tip plastic zone and for any loading condition can be deduced from Eq. 18 and Eq. 19, as:

$$\Delta\varepsilon_{eq}^{vp}(r,\theta)=\underbrace{\left[\Delta\varepsilon_{eq}^{vp}\left(r,\theta=0°\right)\right]}_{\text{obtained by Eq. 17}}{}^{\left(\frac{a\theta^2+b\theta+c}{c}\right)} \tag{20}$$

This type of approach based on the summation of the crack tip viscoplastic strain for each increment of time (Eq. 17 and Eq. 20) intrinsically takes into account all of the temperature variation effect (non-isothermal loading), the time dependent effect (loading frequency), and the effect of the compressive load (for negative loading ratio cycle) on the crack propagation rate.

3.2.2. Monotonic Plastic Zone Size $r_p^{mon}(\theta=0°)$ Calculation

Generally, the monotonic plastic zone size corresponds to the zone is which the maximum equivalent inelastic strain $\varepsilon_{eq\,max}^{vp}$ is greater than 10^{-4}. Thus, the plastic zone size $r_p^{mon}(\theta=0°)$ can be deduced from the asymptotic solution of $\varepsilon_{eq}^{vp}(r,t)$ (cf. Eq. 14) as follows:

$$r_p^{mon}(\theta=0°)=\arg\min_r\left|\varepsilon_{eq}^{vp}\left(r,t_m\right)-10^{-4}\right| \tag{21}$$

Thus, by calculating the monotonic plastic zone size and having already deduced the general form of the viscoplastic strain range, the Eq. 2 can be applied and the crack growth rate can be calculated.

3.2.3. Model Enrichment by Including the History Effect

Under high remote loading, the viscoplastic strain amplitude and the plastic zone size increase, what causes significant history effect due to accumulated plasticity and leads to inaccurate values of stress intensity factor *K*. Thus, the effect of the accumulated plasticity on the crack growth rate has to be taken into account.

To do so, the stress intensity factor *K* used in the previous section to calculate the crack growth rate (Eq. 17, Eq. 20 and Eq. 2), is replaced by a corrected factor K_p defined by:

$$K_{p(i)(l)} = \sqrt{\frac{J_{(i)(l)}E}{1-\nu^2}} \tag{22}$$

where l is the current instant of the cycle i. The term $J_{(i)(l)}$ is determined by summing the elastic part of the J integral obtained in the current cycle i (first term of Eq. 23) and the term J_p^* representing the accumulated inelastic J parts of previous cycles:

$$J_{(i)(l)} = \frac{K_{(i)(l)}^2\left(1-\nu^2\right)}{E} + J_p^* \tag{23}$$

$K_{(i)(l)}$ is the "elastic" stress intensity factor at the instant (l) of the cycle (i). This parameter is calculated using the crack flanks relative displacement (in the area of K dominance) in a coarse mesh finite element model. Consider that the point M corresponds to the current position of the crack tip at cycle (i), the term J_p^* corresponds to the accumulation of the inelastic J at the point M from cycle ($i - j$) to cycle (i). Where ($i - j$) is the cycle for which the crack tip plastic zone reaches the current crack tip position i.e. the point M (cf. Figure 12).

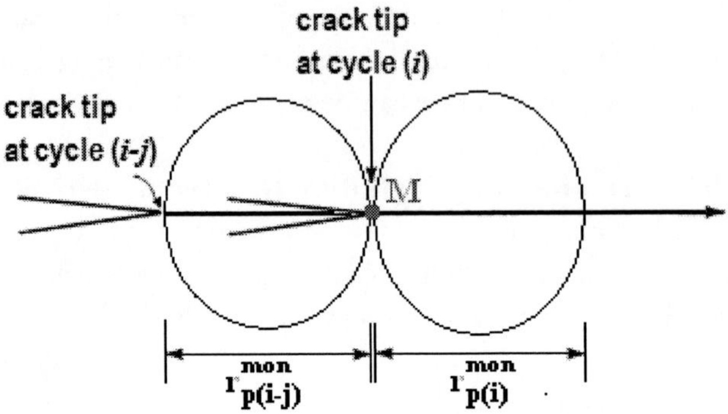

Figure 12. Schematic representation of the crack growth.

It is worth emphasising that, at the point M, the accumulated inelastic strain range grows from cycle $(i-j)$ to cycle (i) due to the approach of the crack tip to point M (decrease of r, cf. Figure 9). Thus, for each cycle between $(i-j)$ and (i), the inelastic J part is calculated by subtracting the elastic part of J integral from its maximum value calculated by finite element ($J_{c\ max}$). The result obtained for each cycle is then multiplied by the term representing the variation of inelastic deformations at point M depending on its distance to the crack tip at a given cycle between $(i-j)$ and (i). Finally, J_{p}^{*} is deduced by summing these last quantities for cycles going from $(i-j)$ to (i):

$$J_{p}^{*}=\sum_{x=\langle i-j\rangle}^{i}\left(J_{\mathrm{c\,max}\,(x)}-\frac{K_{\max(x)}^{2}(1-\nu^{2})}{E}\right)\left(\frac{\Delta\varepsilon_{\mathrm{eq}}^{\mathrm{vp}}\left(r=r_{\mathrm{p}(i-j)}^{\mathrm{mon}}-\sum_{n=i-j}^{x}\Delta a_{(n)}\right)}{\Delta\varepsilon_{\mathrm{eq}}^{\mathrm{vp}}(r\rightarrow 0)}\right) \tag{24}$$

$r_{\mathrm{p}(i-j)}^{\mathrm{mon}}$ is the size of the monotonous plastic zone (at $\theta = 0°$) at cycle $(i-j)$ and Δa is the crack length increase by cycle. $\Delta\varepsilon_{\mathrm{eq}}^{\mathrm{vp}}(r\rightarrow 0)$ is the inelastic strain range calculated at 5 μm from the crack tip at the current position (i). Knowing that $(i-j)$ corresponds to the cycle for which the length of the crack plus the size of the plastic zone are equal to the length of crack at the cycle (i) (cf. Figure 12), cycle number (j) must then verify the following equation:

$$a_{(i-j)}+r_{\mathrm{p}(i-j)}^{\mathrm{mon}}=a_{(i)} \tag{25}$$

By considering, as a first step, that $da/dN_{(i-j)} \approx da/dN_{(i)}$ and that $r^{\mathrm{mon}}_{\mathrm{p}(i-j)} \approx r^{\mathrm{mon}}_{\mathrm{p}(i)}$, a first estimation of the value of (j) is obtained as follows:

$$j = \frac{r^{\mathrm{mon}}_{\mathrm{p}(i)}}{da / dN_{(i)}} \tag{26}$$

Hence, the exact value of (j) is obtained by the following iterative method:

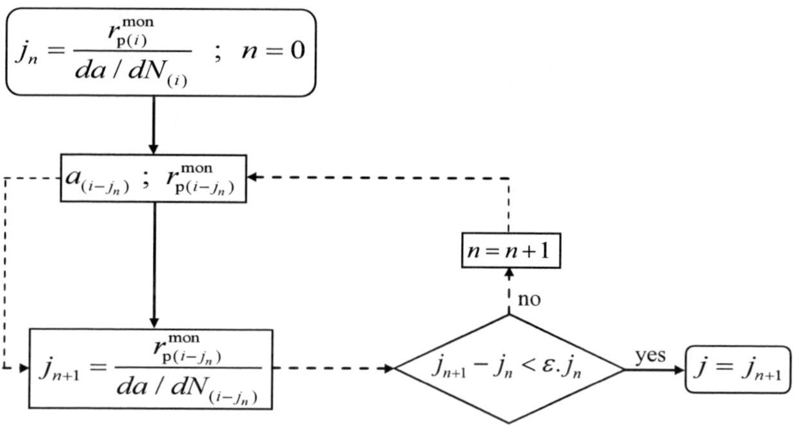

CONCLUSION

Materials of automotive components undergoing thermomechanical fatigue loading are more and more pushed to their limits under increasingly severe safety and pollution standards constraints. Thus, components should be optimally designed for the sake of meeting the industrial requirements while remaining functional and fulfilling their desired lifetime. To do so, a physically based damage tolerant design approach is developed in this work for elasto-viscoplastic cast alloys undergoing thermomechanical fatigue loading. This type of approach is based on an optimal lifetime prediction law that considers both of crack nucleation and crack growth lifetime. Since, most of vehicles components are currently designed using the “safe life” approach (S-N curve) that only considers the crack initiation lifetime, this work has been focused on the crack growth phase. Therefore, thermomechanical fatigue crack growth mechanisms and behaviour have been carefully analysed and understood. These numerical and experimental analyses have shown that the amplitude of the inelastic deformations and the extent of the plastic zone ahead of the crack tip are relevant parameters to describe crack growth mechanisms and to estimate the crack growth rate. Thus, on the bases of these two physical quantities, a new physically based crack growth law has been developed.

The advantage of physically based laws is that they are able to intrinsically take into account the interaction between different damage mechanisms due to fatigue, creep or monotonic loading and that for any loading condition. The law developed in this work consists on the summation of the inelastic strain range in the entire monotonic plastic zone ahead the crack tip using a computational asymptotic solution for an elasto-viscoplastic material. Such a physical approach is capable to consider implicitly all of the purely fatigue effect, monotonic loading effect (time dependent), effect of temperature variation (non-isothermal loading) and effect of the compressive loading. In addition, in order to take into account the history effect, the linear elastic stress intensity factor K was replaced by a corrected factor K_p that accounts the accumulated crack tip plasticity induced by previous cycles. The law, applied for Al-Si alloys, contains only one material parameter (ζ) that may be calibrated experimentally by one simple isothermal fatigue cracking test.

REFERENCES

[1] Elber W. Fatigue crack closure under cyclic tension. *Eng Fract Mech* 1970;2:37–44.

[2] Ritchie RO, Suresh S. Some considerations on fatigue crack closure at near-threshold stress intensities due to fracture surface morphology. *Metal Trans* 1981;13A:937–40.

[3] Suresh S. Crack deflection: implications for the growth of long and short fatigue cracks. *Metal Trans* 1983;14A:2375–85

[4] Newman CJ. A crack-closure model for predicting fatigue crack growth under aircraft spectrum loading. *ASTM STP* 748. 1981; p. 53–84.

[5] Donald K, Paris PC. An evaluation of DKeff estimation procedure on 6061-T6 and 2024-T3 aluminum alloys. *Int J Fatigue* 1999;21:S47–57.

[6] Kujawski D. Enhanced model of partial crack closure for correlation of R-ratio effects in aluminum alloys. I*nt J Fatigue* 2001;23:95–102.

[7] McEvily AJ, Ishihara S, Endo M, Sakai H, Matsunaga H. On one- and two-parameter analyses of short fatigue crack growth. *Int J Fatigue* 29 (2007) 2237-2245.

[8] McEvily AJ, Endo M, Yamashita K, Ishihara S, Matsunaga H. Fatigue notch sensitivity and the notch size effect. *Int J Fatigue* 30 (2008) 2087-2093.

[9] Kujawski D. A new (DK+Kmax)0.5 driving force parameter for crack growth in aluminum alloys. *Int J Fatigue* 2001;23:733–40.

[10] Kujawski D. A fatigue crack driving force parameter with load ratio effects. *Int J Fatigue* 2001;23:S239–46.

[11] Dinda S, Kujawski D. Correlation and prediction of fatigue crack growth for different R-ratios using Kmax and DK+ parameters. *Eng Fract Mech* 2004;71:1779–90.

[12] Huang X, Moan T. Improved modeling of the effect of R-ratio on crack growth rate. *Int J Fatigue* 29 (2007) 591-602.

[13] Reinhart TL. ASM Handbook. vol. 19. ASM International. *Materials Park.* OH. USA. 1996. pp. 813-822.

[14] McLellan DL, McLellan M.M. An approach to fatigue design for aluminum castings. *AFS Trans.* 104 (1996) 445-449.

[15] Major JF. Porosity control and fatigue behaviour in A356T61 aluminum alloy. *AFS Trans.* 105 (1997) 901-906.

[16] Wang QG, Apelian D, Griffith JR. Microstructural effects on the fatigue properties of aluminum castings. in: M. Tiryakioglou. J. Campbell (Eds.). *Proceedings from Materials Solutions Conference'98 on Advances in Aluminum Casting Technology.* 12-15 October 1998. pp. 217-223.

[17] Wang QG, Apelian D, Lados DA. Fatigue behavior of A356/357 aluminum cast alloys. Part II. Effect of microstructural constituents. *J Light Met* 2001;1(1):85–97.

[18] Topper TH, Au P. The effects of stress ratio. compressive load and underload on the thresold behaviour of a 2024-T351 alloy. In *2nd International Conference On Fat.* And Fta. Thresholds. Brimingham. UK: Chamelon Press; 3–17 September 1984. 179–90.

[19] Tack AJ, Beevers CJ. The influence of compressive loading on fatigue crack propagation in three aerospace bearing steels. In: *4th International conference on fat. and fat.* thresholds. UK: MPCE Ltd; Honolulu. 15–20 July 1990. p. 1179–84.

[20] Pommier S, Prioul C, Bompard P. Influence of a negative R ratio on the creep fatigue behaviour of the N18 nickel base superalloy. *Fatigue Fract Eng Mater Struct* 1997;20(1):93 107.

[21] Lados DA, Apelian D, and Major JF. Fatigue crack growth mechanisms at the microstructure scale in Al–Si–Mg cast alloys: mechanisms in Regions II and III. *Metall Mater Trans* A;37A(8):2405–18. 2006.

[22] Lados DA, Apelian D. Relationship between microstructure and fatigue crack propagation paths in Al-Si-Mg cast alloys. *Engng Fract Mech:*75(3-4):821-832. 2008.

[23] Pernot JJ, Nicholas T, Mall S. Modelling thermomechanical fatigue crack growth rates in Ti-24AL-11Nb. *Butterworth-Heinemann Ltd* 16; 1994:111-122.

[24] Jordan EH, Meyers GJ. Fracture mechanics applied to nonisothermal fatigue crack growth. *Eng. Fract. Mech*. 23(2);1986:345-358.

[25] Pommier S. Cyclic plasticity and variable amplitude fatigue. *Int J Fatigue* 2003;25:983–97.

[26] Silva FS. Crack closure inadequacy at negative stress ratios. *Int J Fatigue* 2004;26:241–52.

[27] Romeiro FFJ, Domingos CA, Freitas MJM. *Measurement of fatigue crack closure for negative stress ratio. Advances in fatigue crack closure measurement and analysis.* STP 1343. Philadelphia (PA):ASTM; 1997. p. 321–36.

[28] Fleck NA, Shin CS, Smith RA. Fatigue crack growth under compressive loading. *Eng Fract Mech* 1985;21(1):173–85.

[29] Suresh S. Crack initiation in cyclic compression and its applications. *Eng Fract Mech* 1985;21(3):453–63.

[30] Pippan R. The growth of short cracks under cyclic compression. *Fatigue Fract Eng Mater Struct* 1987;9(5):319–28.

[31] Hermann R. Fatigue crack growth in ductile materials under cyclic compressive loading. *Fatigue Fract Eng Mater Struct* 1994;17(1): 93–103.

[32] Kasaba K, Sano T, Kudo S, Shoji T, Katagiri K, Sato T. Fatigue crack growth under compressive loading. *J Nucl Mater* 1998;258–263:2059–63.

[33] Lemaitre J, Chaboche JL. *Mechanics of solid materials.* Cambridge University Press (1990).

[34] Merhy E, Remy L, Maitournam H, Augustins L. Crack growth characterisation of A356-T7 aluminium alloy under thermo-mechanical fatigue loading. *Engng Fract Mech* (2013).

[35] Maitournam MH, Pommier B, Thomas JJ. Determination of the asymptotic response of a structure under cyclic thermomechanical loading. *Comptes Rendus Mecanique* 330 (2002), 10:703-708.
[36] McClintock FA. Fracture of solids, pp. 65-102. Interscience, New York, 1963.
[37] Chalant G, Remy L. Model of fatigue crack propagation by damage accumulation at the crack tip. *Engineering Fracture Mechanics* 18(5);1983:939–952.
[38] Sih GC, Jeong DY. Fatigue load sequence effect ranked by critical available energy density. *Theoretical and Applied Fracture Mechanics* 14;1990:141–51.
[39] Wang W, Hsu CTT. Fatigue crack growth rate of metal by plasticenergy damage accumulation theory. *Journal of Engineering Mechanics* 120(4);1994:776–95.
[40] Skelton RP, Vilhelmsen T, Webster GA. Energy criteria and cumulative damage during fatigue crack growth. *Int. J. Fatigue,* 20(9);1998:641-649.
[41] Klingbeil NW. A total dissipated energy theory of fatigue crack growth in ductile solids. *Int J Fatigue* 2003;25:117–128.
[42] Birol Y. What happens to the energy input during fatigue crack propagation. *Materials Science and Engineering* 1988;A104:117–24.
[43] Stuwe HP, Pippan R. On the energy balance of fatigue crack growth. *Computers and Structures* 44(1–2);1992:13–7.
[44] Ranganathan N, Desforges JR. Equivalent constant amplitude concepts examined under fatigue crack propagation by block loading. *Fatigue and Fracture of Engineering Materials and Structures* 19(8);1996:997–1008.
[45] Rajic NL, Lam YC, Wong AK. Evolved heat as a fatigue characterisation parameter. *Materials Science Forum* 1996: 210-213,463–70.
[46] Hutchinson JW. Singular behavior at the end of a tensile crack in a hardening material. *Journal of the Mechanics and Physics of Solids,* 16;1968:13–31.
[47] Rice JR. A path independent integral and the approximate analysis of strain concentrations by notches and cracks. *Journal of Applied Mechanics* 35;1968:379–386.
[48] Hoff NJ. *Quarterly of Applied Mechanics* 12;1954:49-55.
[49] Riedel H, Rice JR. Tensile cracks in creeping solids. *Fracture Mechanics: Twelfth Conference*, ASTM STP 700;1980:112-130.

[50] Shih CF. Tables of Hutchinson-Rice-Rosengren singular field quantities. *MRL E-147,* Division of Engineering, Brown University, 1983.

[51] Chao YJ, Zhang L. Tables of plane strain crack tip fields: HRR and higher order terms, *Me-report 97-1*, Department of Mechanical Engineering, University of South California, 1997.

[52] Bassani JL, McClintock FA. Creep relaxation of stress around a crack tip. *International Journal of Solids and Structures* 17, 1981:479-492.

[53] Rice JR. Mechanics of Crack Tip Deformation and Extension by Fatigue. *In Fatigue Crack Propagation - ASTM STP 415*; 1967:247–309.

[54] Creager M, Paris PC. *International Journal of Fracture Mechanics* 3;1967:247-252.

[55] Pinho de Castro JT, Meggiolaro MA, Miranda AC. Fatigue crack growth predictions based on damage accumulation calculations ahead of the crack tip. *Computational Materials Science*, 46;2009:115-123.

In: Recent Trends in Fatigue Design
Editor: Ricardo Branco

ISBN: 978-1-61668-410-5

Chapter 5

RECENT DEVELOPMENT OF FATIGUE CRACK GROWTH RATE MODELS

Fang Wang [*] ***and Weicheng Cui***
Hadal Science and Technology Research Center, Shanghai Ocean University, China

ABSTRACT

The crack growth rate curve is the fundamental material resistance property to fatigue crack growth. Up to now, hundreds of empirical models have been proposed for the description of the fatigue crack growth rate curve and many recent progresses have been made to explain different fatigue phenomena such as environmental effects, structural effects, loading effects and material effects. Most of them are reversal based with cycle-counting but some new models aim at expressing fatigue crack growth at the small time scale based on the incremental crack growth at any time instant within one cycle. To explain the difference between crack initiation and growth regime, the models can be classified into small crack growth rate model and large crack growth rate model. And from the perspective of driving force controlling parameters, there are single-parameter models and two-parameter models. To explain the interaction with other damage modes, some models are modified considering effects such as creep. In this chapter, the recent development

[*] Corresponding Author address. Email: wf0224@aliyun.com.

of fatigue crack growth rate models are reviewed and some of their applications are demonstrated.

Keywords: Crack growth rate model, reversal based model, small-scale model, creep effect

1. INTRODUCTION

Fatigue cracks usually initiate from the surface of a component, where fatigue damage initiates as shear cracks on crystallographic slip planes. The surface shows the slip planes as intrusions and extrusions. This is Stage I crack growth. After a transient period, Stage II crack growth takes place in a direction normal to the applied stress. Finally the crack becomes unstable and fracture occurs. Figure 1 shows a schematic representation of the two stages (Cui, 2002).

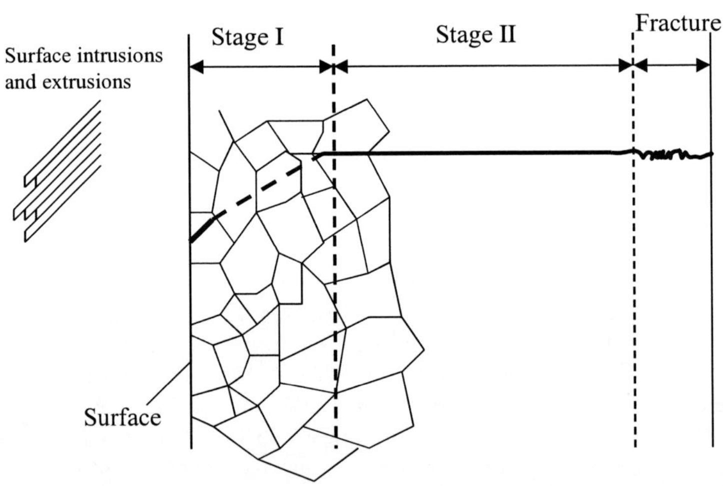

Figure 1. Schematic representation of crack formation and growth in polycrystalline metals (Cui, 2002).

This two-stage process was first recognized by PJE Forsyth (1969) and it was one of the most important achievements in metal fatigue in the 20th century (Miller, 1999). But that division may be too rough for further discussion. So in some other sources, this process may be further divided. It is recommended in Cui (2002) that the total process of fatigue failure is divided into five stages: (1) crack nucleation ($a<a_m$); (2) microstructurally small crack propagation ($a_m<a\leq a_p$); (3) physically small crack propagation ($a_p<a\leq a_l$); (4) long crack propagation ($a_l<a\leq a_c$); (5) final fracture; where a is the characteristic dimension of an equivalent crack in a component, a_m is the smallest crack length detectable by current technology which is about 0.1μm; a_p is the smallest crack length for physically small cracks which is about 10μm, a_l is the smallest crack length for long cracks which is about 1 mm, a_c is the critical crack length at which component fracture occurs. Obviously, these boundary divisions are subjective and depends on the ability of crack measuring systems.

Traditionally, the processes before long crack propagation is named "fatigue crack initiation" while long crack propagation is called "fatigue crack propagation". This division is also too rough and could not be expected to provide an accurate prediction of fatigue life. However, it may be worth to point out here that depending on the initial crack length in a component, some of the early stages may be skipped. These five stages of fatigue process only exists in "defect free" metal components. In this book, emphases are put on physically small fatigue cracks (short fatigue cracks) and long fatigue cracks (large cracks).

The different behavior of small fatigue cracks and long fatigue cracks has been found for several decades of years. The small- or short-crack growth characteristics are affected by many influential factors such as loading conditions together with the metallurgical features. Experiments at relatively high stress levels show that the concepts based on linear elastic fracture mechanics (LEFM) could not be used to explain the observed small-crack effects, because the plastic zone sizes would be large compared to the crack sizes. Then nonlinear fracture mechanics (elastic-plastic fracture mechanics) can be used. At lower stress-intensity, small fatigue cracks are found to grow faster than predicted from large crack data using the models based on LEFM. The phenomenon can be illustrated in Figure 2 (Ritchie, 1986). Where, the crack-growth rate is plotted against the stress intensity factor range, ΔK. Typical large-crack results (solid curve) are obtained from tests with cracks greater than about 2 mm in length. The large crack threshold, ΔK_{th}, is usually obtained from load-reduction (ΔK-decreasing) tests.

Typical results for small cracks at various applied stress levels (Si) are shown by the dashed curves. Small fatigue cracks initiate at inclusion particles, voids, or weak grains do not have any prior plastic deformation to develop closure. Thus, the crack-opening stress behavior must go through a transient behavior to reach steady-state conditions. (A steady-state crack is one in which residual-plastic deformations and crack-opening stresses have stabilized under constant-amplitude loading conditions.) If a small crack is fully open, then the stress-intensity factor range is fully effective and the crack-growth rate is faster than the steady-state crack-growth rate. Because the large-crack threshold may also be elevated due to load history effects, the steady-state crack-growth rate behavior may lie between the small-crack and large-crack-threshold behavior, as shown by the dash-dot curve in Figure 2.

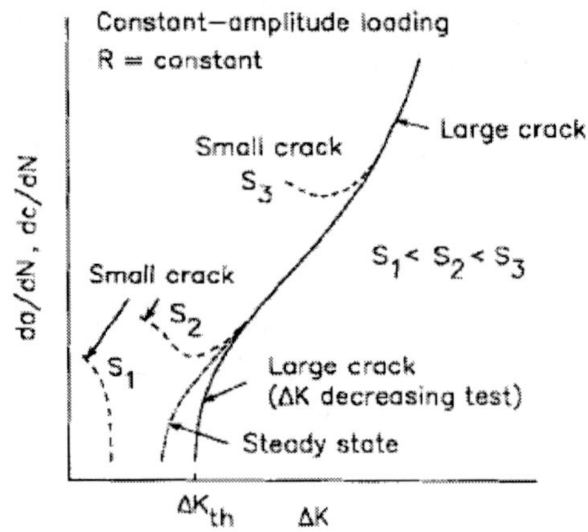

Figure 2. Typical fatigue-crack-growth rates for large cracks and small cracks at various applied stress (Ritchie, 1986).

In this chapter, the small crack growth rate models and large crack growth rate models are introduced including single-parameter reveral based models, two-parameter reversal based models and a time-domain small-scale model. To explain the interaction with other damage modes, some models are modified considering effects of creep.

2. LONG CRACK GROWTH RATE MODEL

2.1. Characteristics of Long Crack Growth

(1) Sigmoidal Shape

The fitted curve for test data of long crack growth is shown in Figure 3, which is a sigmoidal curve. Three regions can be divided in the curve including the initiation and threshold region where the crack growth rate goes asymptotically to zero as ΔK approaches a threshold value Δ_{th} and there is no crack grow th below Δh_{th} such as the highlighed test data shown in Figure 4, stable crack growth region where crack growth follows a straight line, and unstable fracture region where a rapidly increasing growth rate towards "infinity", i.e. ductile tearing and/or brittle fracture occurs.

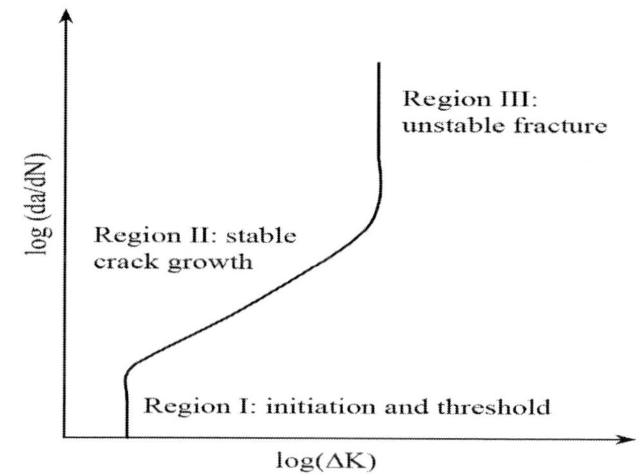

Figure 3. The sigmoidal shape of crack growth.

(2) Load Ratio Effect

It is observed that the value of stress ratio $R(=\sigma_{min}/\sigma_{max})$ and mean stress affect the growth in a manner that increasing R increases the growth rate and vise versa. Then the stress ratio effect is sometimes called the mean stress effect, especially in the approaches based on CFD theory. As an example, some data illustrating R ratio effect on long crack growth rates for an titanium alloy Ti80 is shown in Figure 5. An approach to the effect of mean stress on the threshold conditions, i.e. fatigue limit and threshold stress intensity factor range, is presented by Kujawski and Ellyin (1995). The approach is based on the premise that damage caused by cyclic loading is a function of the strain energy range ΔW, i.e. a function of both stress and strain ranges. It is shown that the mean stress effect on the threshold conditions can be treated in a unified manner for members with and without crack-like flaws.

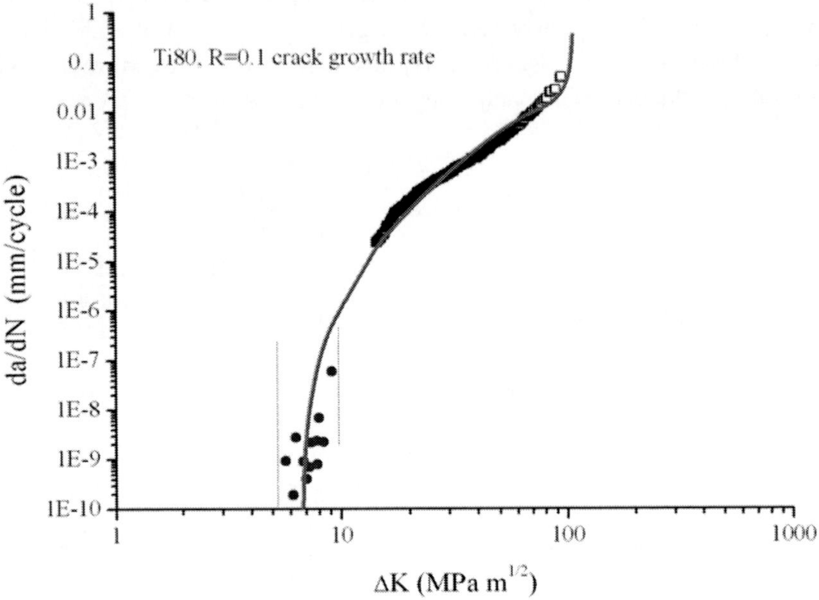

Figure 4. The illustration of crack growth near threshold for an titanium alloy Ti80.

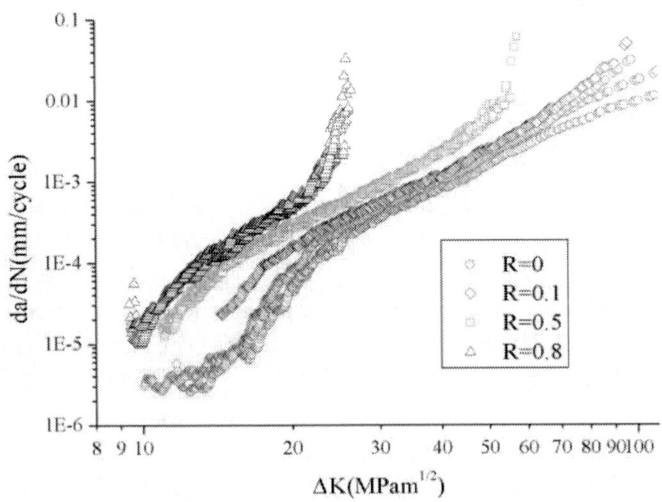

Figure 5. Effect of R ratio on long crack growth rates for an titanium alloy Ti80.

(3) Overload Effect

Significant load interaction effect can occur as a result of variable loading history is shown in Figure 6. Crack growth rate is essentially affected by three load sequences: (i) a tensile overload, (ii) a compressive underload, (iii) a tensile overload followed by a compressive underload, which are commonly referred to as crack retardation, crack acceleration and reduction in crack retardation, respectively (Rushton and Taheri, 2003). The fatigue crack growth behaviors under random loading histories such as Flight Spectra are usually investigated as the combination of the crack growth behaviors under three basic load sequences mentioned above. In literature, more attention are paid on overloading effects as crack growth retardation due to overload is the most commonly observed load interaction effect and has the most notable influence on fatigue life. Overload effect is typically explained by crack closure theory. Immediately after the overload, there will be an instant increase of crack growth higher than the baseline growth rate and an overload plastic zone occurs in the vicinity of crack tip. The crack opening level instantaneously rises to the maximum value, and then gradually recovers to the initial level under constant amplitude loading when the crack penetrates the large plastic zone in the subsequent cycles. Then there would be a number of delay cycles in the post-overload growth. Furthermore, it is observed that the number of delay cycles is affected by load ratio, see Figure 6.

(4) Thickness Effect

Variable-amplitude fatigue studies of 2024-T3 aluminum alloy were performed to examine the effect of sheet thickness on fatigue crack growth rate retardation. Results indicated that the amount of retardation increased with decreasing specimen thickness. This phenomenon was attributed to enhanced plastic strains under plane stress conditions (i.e. in a thin sheet) which formed ahead of the advancing crack tip as a result of a high load excursion. These strains are believed to produce both crack closure and a favorable compressive residual stress field around the crack tip. Evidence of increased crack surface interference under plane stress situations was verified with electron fractographic observations. Experimental validation of thickness effect can be shown in Figure 7 (Bao and McEvily,1998).

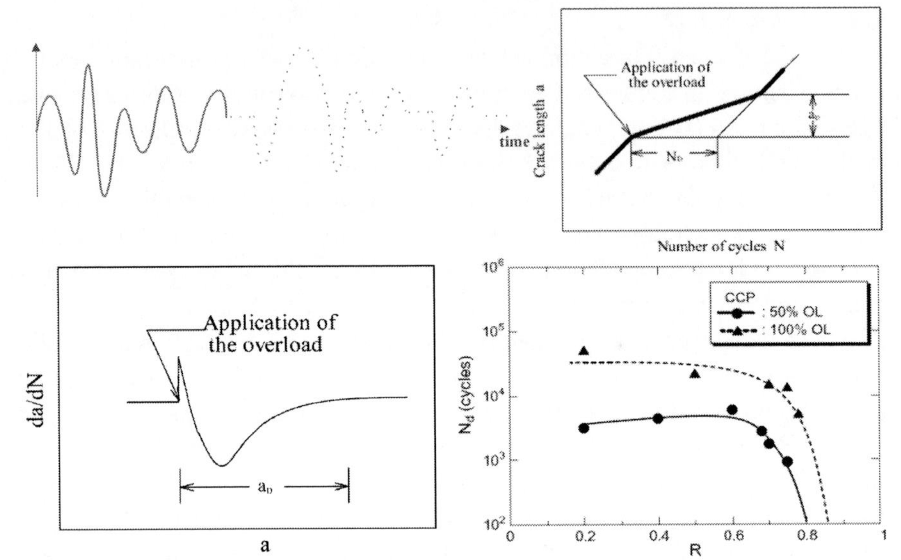

Figure 6. The illustration on random loading history for engineering structures; the defination of delay cycles; characteristic features of the variation in the post-overload fatigue crack growth rate (McEvily and Yang,1990); and the number of delay cycles as a function of R for a center-cracked panel of 6061-T6 (Ishihara et al., 2008).

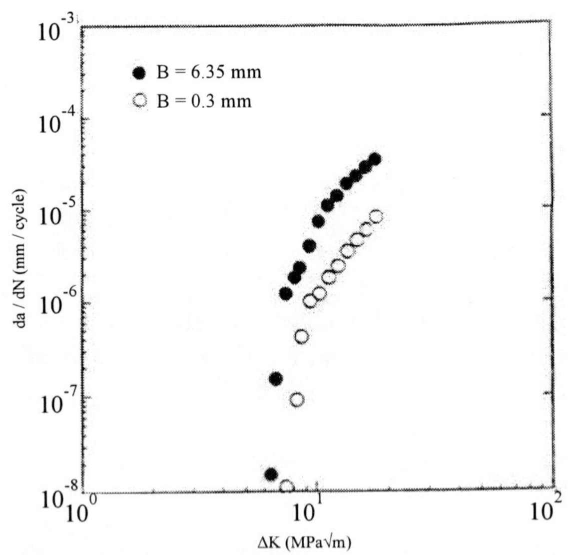

Figure 7. Thickness effect (Bao and McEvily, 1998).

2.2. Reversal Based Models

Typically, the crack growth behavior can be described the relationship between cyclic crack growth rate da/dN and stress intensity range ΔK based on the linear elastic fracture mechanics (LEFM) approach, which is called the reversal based model.

In the early stage of establishing crack growth rate model, the second region is noticed. The first model was Paris Law (1961), which represents the relationship by a straight line on the log-log plot and still widely used in engineering,

$$\frac{da}{dN} = A(\Delta K)^n \tag{1}$$

Where A and n are material constants; n is the slope on the log-log plot assuming that the decades on both log scales are the same length; and Y is a geometry factor depending on the loading and cracked body configuration.

Some time later, when people found the nonlinearity for all the ranges of ΔK. Donahue et al (1972) proposed the crack growth relation considering the threshold region:

$$da/dN = C \cdot [\Delta K - \Delta K_{th}]^m \tag{2}$$

Later on, Forman et al (1967) proposed the crack growth expressing the total sigmoidal shape of crack growth with R ratio:

$$\frac{da}{dN} = \frac{C(\Delta K)^m}{(1-R)K_c - \Delta K} \tag{3}$$

where K_c is the fracture toughness of the material.

From the sixties of last century, lots of long crack growth rate models have been proposed and each has its advantages and drawbacks (e.g. Cui et al, 2011). Most of them can combine the behaviors at high, intermediate and low ΔK values. Recently, more progress has been made to establish crack growth rate relations which are able to explain different fatigue phenomena. According to the threshold parameter used in those models, they can be subdivided into two catergories as "single-parameter model" which illustrates one threshold value ΔK_{th} as the early models, and "two-parameter model" with two threshold parameters ΔK_{th} and K_{maxth} in which it is considered that the crack will propagate only when ΔK and K_{max} reach their threshold values in the meanwhile. Several typical models in the two catergories will be briefly introduced in the following text.

As one of the fatigue crack growth models in the single-parameter models, the modified constitutive relation developed by McEvily and his co-workers (McEvily and Ishihara, 1999) can not only account for the effects of initial crack size and load sequence, but also explain various other phenomena of metal fatigue observed in tests, which has been proved continuously recently

But, the modified model are only for linear and near threshold regions and valid for elastic-perfectly-plastic materials. Consequently, this modified constitutive relation is further generalized by Cui and Huang (2009) in three main aspects: (1) introducing an unstable fracture condition; (2) defining a virtual strength to replace the yield stress; and (3) defining an overload and underload parameter. The performances of this general constitutive relation for fatigue crack growth have been extensively studied and it is found that this general constitutive relation is able to explain various phenomena observed with particular strong capability on load sequence effect. Based on the work of Cui and Huang (2009), a further extension is made by Wang et al (2008) concerning two aspects: (1) the slope of the fatigue crack growth rate curve is regarded as a variable rather than a fixed value for different materials; (2) both the maximum stress intensity factor at the crack opening level and the effective stress intensity factor range at the threshold level are written as the functions of load ratio.

$$\frac{da}{dN} = AM^2 \tag{4}$$

$$M = K_{\max}(1-R) - \left(1-e^{-ka}\right)\left(K_{op\max} - RK_{\max}\right) - \Delta K_{effth} \tag{5}$$

$$K_{\max} = \sqrt{\pi r_e \left(Sec\frac{\pi}{2}\frac{\sigma_{\max}}{\sigma_Y} + 1 \right)} \cdot \left(1 + Y(a)\sqrt{\frac{a}{2r_e}} \right) \sigma_{\max} \tag{6}$$

where r_e is the size of an inherent flaw, a parameter whose magnitude is of the order of several microns in length. Newly formed cracks are only significant when their lengths exceed r_e. a is the modified crack length which is equal to

r_e plus the actual crack length. Thus the minimum value of a is r_e, which represents a component containing no crack (plain specimen). k is a material constant which reflects the rate of crack closure development with crack advance and is similar to that in other partial crack closure models (Kujawski, 2001). Y is a geometrical factor. K_{opmax} is the maximum stress intensity factor at the opening level for a macroscopic crack, $MPam^{0.5}$. Both K_{opmax} and ΔK_{effth} are functions of R. This modified constitutive relation was found to be able to account for various fatigue phenomena observed: (1) anomalous fatigue crack growth; (2) fatigue crack growth under compression-compression cycling; (3) delay due to an overload; (4) two step cyclic loading; (5) the effect of mean stress on fatigue life; (6) small fatigue crack growth. Especially, it is valid from physically small crack to long crack regimes (McEvily and Ishihara, 2001), which will be further introduced in Section 3.

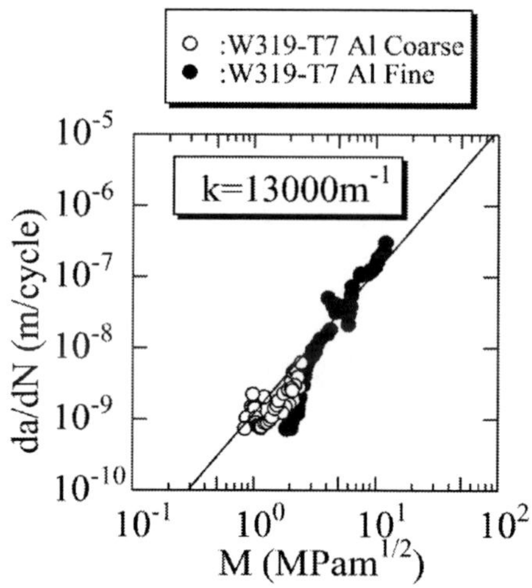

Figure 8. The rate of fatigue crack growth of small cracks in the aluminum alloy W319-T7 as a function of the parameter M (Ishihara & McEvily, 2002).

However, there are still some shortcomings in these empirical formulas, for example, the ability of the model to explain some critical phenomena as overload retardation and underload acceleration, specimen thickness effect, the unstable fracture condition etc. In the past several years, the authors' group have made some efforts in improving the ability of the crack growth rate model in UFLP for marine structures. The extended McEvily model was further improved by Yi et al. (2008), Wang & Cui (2010), Chen & Cui (2012).

And validation on the improved model has been made by comparing the predicted results with test data for a wide range of materials.

$$\begin{cases} da/dN = \dfrac{A\left[K_{\max}\cdot(1-\phi f_{\mathrm{op}})-\Delta K_{\mathrm{effth}}\right]^m}{1-(K_{\max}/K_{\mathrm{Cf}})^n} \\ K_{\max} = \sqrt{\pi r_{\mathrm{e}}\left(\mathrm{Sec}\dfrac{\pi}{2}\dfrac{\sigma_{\max}}{\sigma_{\mathrm{v}}}+1\right)}\left(1+Y(a)\sqrt{\dfrac{a}{2r_{\mathrm{e}}}}\right)\sigma_{\max} \\ K_{\min} = \sqrt{\pi r_{\mathrm{e}}\left(Sec\dfrac{\pi}{2}\dfrac{\sigma_{\min}}{\sigma_{\mathrm{v}}}+1\right)}\left(1+Y(a)\sqrt{\dfrac{a}{2r_{\mathrm{e}}}}\right)\sigma_{\min} \\ \Delta K = K_{\max}-K_{\min} \\ \dfrac{\sigma_{\mathrm{v}}}{\sigma_{\mathrm{u}}} = \dfrac{\pi}{2}\cdot\dfrac{1}{\cos^{-1}(\frac{1}{\alpha^2-1})},\quad \alpha = \dfrac{K_{\mathrm{C}}}{\sigma_{\mathrm{u}}\sqrt{\pi r_{\mathrm{e}}}\left(1+\frac{Y(r_{\mathrm{e}})}{\sqrt{2}}\right)} > \sqrt{2} \end{cases} \tag{7}$$

where,

$$f_{\mathrm{op}} = \begin{cases} \max\{R, A_0+A_1R+A_2R^2+A_3R^3\} & 0 \le R < 1 \\ A_0+A_1R & -2 \le R < 0 \end{cases}$$

$$\begin{cases} A_0 = (0.825-0.34\alpha'+0.05\alpha'^2)\cdot[\cos(\pi\sigma_{\max}/2\sigma_{\mathrm{fl}})]^{1/\alpha'} \\ A_1 = (0.415-0.071\alpha')\cdot\sigma_{\max}/\sigma_{\mathrm{fl}} \\ A_2 = 1-A_0-A_1-A_3 \\ A_3 = 2A_0+A_1-1 \\ \sigma_{\mathrm{fl}} = (\sigma_{\mathrm{Y}}+\sigma_{\mathrm{u}})/2 \\ \alpha' = \dfrac{1}{1-2\nu}+\dfrac{1-\dfrac{1}{1-2\nu}}{[1+0.8861\cdot(t/(K_{\max}/\sigma_{\mathrm{Y}})^2)^{3.2251}]^{0.75952}} \end{cases} \tag{8}$$

$$\begin{cases} K_{\mathrm{Cf}} = \left[\dfrac{(1-2\nu)^2-\sqrt{1-\nu^2}}{(1-2\nu)^2-1}\cdot\dfrac{\pi\cdot\lambda}{(1-2\nu)^2}+\dfrac{\sqrt{1-\nu^2}-1}{(1-2\nu)^2-1}\right]\cdot K_{\mathrm{IC}} \\ \lambda = \dfrac{(1\text{-}1.65\nu)^2}{5}-\dfrac{1}{20n'}((1\text{-}1.65\nu)^2)^{\frac{1}{n'}}+\dfrac{\frac{1}{\pi}-\frac{1}{2.2n'}(\frac{1}{\pi})^{\frac{1}{n'}}-(\frac{(1\text{-}1.65\nu)^2}{5}-\frac{1}{20n'}((1\text{-}1.65\nu)^2)^{\frac{1}{n'}})}{(1+\frac{t/(K_{\max}/\sigma_{\mathrm{Y}})^2}{1+1/n'})^{1.6+1/n'}} \end{cases} \tag{9}$$

$$\begin{cases} \text{Overload:}\phi = \begin{cases} \left(\dfrac{a_{\mathrm{OL}} + r_{\mathrm{OL}} - r_{yi}}{a_i}\right)^{\gamma} & a_{\mathrm{OL}} \le a_i \le a_{\mathrm{OL}} + r_{\mathrm{OL}} - r_{yi} \\ 1 & a_i > a_{\mathrm{OL}} + r_{\mathrm{OL}} - r_{yi} \end{cases} \\ \text{Underload:}\phi = \begin{cases} \left(\dfrac{a_i + r_{yi} - r_{\mathrm{UL}}}{a_{\mathrm{UL}}}\right)^{\gamma} & a_{\mathrm{UL}} \le a_i \le a_{\mathrm{UL}} + r_{\mathrm{UL}} - r_{yi} \\ 1 & a_i > a_{\mathrm{UL}} + r_{\mathrm{UL}} - r_{yi} \end{cases} \\ \text{Overload-underload combination:}\phi = \begin{cases} \left(\dfrac{a_{\mathrm{OL}} + r_{\mathrm{OL}} - r_{\mathrm{yi}} - r_{UL}}{a_i}\right)^{\gamma} & a_{\mathrm{OL}} \le a_i \le a_{\mathrm{OL}} + r_{\mathrm{OL}} - r_{\mathrm{yi}} - r_{UL} \\ 1 & a_i > a_{\mathrm{OL}} + r_{\mathrm{OL}} - r_{\mathrm{yi}} - r_{UL} \end{cases} \end{cases} \tag{10}$$

where A is a material- and environmentally-sensitive constant of dimensions $(\mathrm{MPa})^{-2}$; m is a constant representing the slope of the corresponding fatigue crack growth rate curve; n is the index indicating the unstable fracture; K_C is the plane stress fracture toughness of the material; K_{Ic} is the plan stain fracture toughness of the material; K_{cf} is the fracture toughness of the material under fatigue loading; r_e is an empirical material constant of the inherent flaw length of the order of 1 μm; a is the modified crack length which is equal to r_e plus the actual crack length; σ_{max} is the maximum applied stress, σ_{min} is the minimum applied stress; Y(a) is a geometrical factor; Y(r_e) is a geometrical factor when a is equal to r_e; R is the stress ratio; ΔK_{eff} is the effective range of the stress intensity factor; ΔK_{effth} is the effective range of the stress intensity factor at the threshold level; K_{op} is the stress intensity factor at the opening level; α' is the crack tip stress/strain constraint ratio, which is 1 for the plane stress state and 1/(1-2ν) for the plane strain state; σ_u is the ultimate strength of the material; σ_Y is the yield strength of the material; n' is the hardening exponent of the power-law material; ν is the Poisson's ratio. It can be noticed that K_{max}, f_{op} and K_{cf} are all the functions of crack length a. γ is a shaping exponent determined by fitting the recovering period of the experimental data. a_i is the crack length in the current ith cycle; r_{yi} is the plastic zone in the current ith cycle; a_{OL} is the crack length when overload cycle occurs; r_{OL} is the plastic zone size when overload cycle occurs; r_{UL} is the plastic zone size when underload cycle occurs. The effect of n is significant only in the unstable propagation region; a constant value of 6 is recommended for a quick and simple engineering analysis.

In recent years, Sadananda and Vasudevan (2003) criticized the fatigue crack growth methods based on the crack closure concept. They pointed out that the drawbacks in the current fatigue life prediction methods stem from several sources: (1) the assumption of plasticity induced crack closure, (2) the lack of terms in the model that relate to the environmental effects and slip deformation behavior, and (3) several adjustable parameters needed to fit the observed data. In order to overcome these drawbacks and develop a more reliable fatigue life prediction model, Vasudevan, Sadananda and co-workers have proposed a two-parameter unified approach.

The crack growth rate expression for the two parameter model has not been explicitly given in many of their published papers. However, Vasudevan et al. (1994) wrote it as

$$\frac{da}{dN} = A\left(\Delta K - \Delta K_{th}^{*}\right)^{m}\left(K_{\max} - K_{\max th}^{*}\right)^{p} \tag{11}$$

and Sadananda and Vasudevan (1997) expressed it as

$$\frac{da}{dN} = f\left[\left(\Delta K - \Delta K_{th}^{*}\right),\left(K_{\max} - K_{\max th}^{*}\right)\right] \tag{12}$$

These two can only cover near threshold and linear regions but cannot cover the unstable fracture condition.

In order to describe the sigmoidal shape of the crack growth rate curve, Bukkapatnam and Sadananda (2005) proposed the following expression:

$$\begin{cases} \dfrac{da}{dN} = C_1 \Psi(\Delta K, K_{\max}) \\ \Psi(\Delta K, K_{\max}) = (s_0 - 2\gamma(\Delta K, \Delta K_0, K_{\max}))K_{\max}^{n} \\ \gamma(\Delta K, \Delta K_0, K_{\max}) = \dfrac{e^{b(\Delta K_0 - \Delta K)^m}}{e^{b(\Delta K_0 - \Delta K)^m} + e^{b(\alpha_1 K_{\max} + \alpha_2 \Delta K + \alpha_3 \Delta K_0)^m}} \end{cases} \tag{13}$$

Here, b determines the flatness of the cusp of the sigmoidal function, m is a real number between 0 and 1, and n is a real number between 1 and 5, ΔK_0 is the value of ΔK and s_0 is related to the value of da/dN at which the curve reaches the cusp before an accelerated fatigue growth takes place, and α_1, α_2, and α_3 determine the relative contributions of ΔK and $K_{\max}$. These parameters can be adjusted to accurately capture the relative influences of these two drivers. As stated in the foregoing, this structure offers flexibility in capturing the complicated fatigue crack growth profiles prevailing in certain material systems, as well as the multifarious environment-fatigue interactions. Furthermore, the structure can correctly capture the relative contributions of the two drivers ΔK and K_{max}, under various loading conditions and environments.

Although their expression can capture the sigmoidal shape, the physical meanings of the parameters are unclear. We recommend writing the general expression of the two-parameter unified approach following the same idea as McEvily and Groeger (1977) as follows (Cui, Bian, Liu, 2007):

$$\frac{da}{dN} = \frac{A\left(\Delta K - \Delta K_{th}^{*}\right)^{m}\left(K_{max} - K_{max\,th}^{*}\right)^{p}}{1-\left(\frac{K_{max}}{K_{cf}}\right)^{n}} \tag{14}$$

This expression has been found to be able to explain many fatigue phenomena observed (Bian et al. 2009; Bian et al. 2010a; 2010b), especially the stress ratio effect.

2.3. Small Scale Crack Growth Rate Model in Time Domain

Almost all existing fatigue prediction models are reversal-based approach (or cycle-based formulation) such as S-N curve, ε-N curve, da/dN equation. However, for most of engineering structures, load history is usually very complex and far from being regularly cycled as shown in Figure 6. The converting from random load history to the cyclic load sequence (such as cycle counting technique) introduces additional complexities and uncertainties in the fatigue damage modeling. And the cycle-based approach makes impossible to continue reducing the time scale for more fundamental investigation since the smallest time scale is one cycle (Lu and Liu, 2010).

In order to overcome these deficiencies of the cycle-based formulations, a new fatigue crack growth formulation based on the small time scale crack increment is proposed by Liu and his co-workers in recent years (Lu and Liu, 2010). The key idea is to define the fatigue crack kinetics at any arbitrary time instant (t+dt) within a load cycle. Their model is introduced as follows.

A schematic illustration is shown in Figure 9. The crack will extend a distance, da, after a small temporal increment, dt, during loading. The model is developed based on the geometric relationship between the Crack Tip Opening Displacement (CTOD) and the instantaneous crack growth kinetics, which is schematically shown in Figure 10. Considering the geometry of crack tips at two time instants (t and t+dt), when the crack tip propagate from point O to O', the CTOD grows from δ to δ', the relationship between the CTOD increment dδ and the crack growth da can be expressed as

$$da = \frac{ctg\theta}{2} d\delta = C d\delta \tag{15}$$

where $C = \frac{ctg\theta}{2}$ and θ is the crack tip opening angle (CTOA). Eq. 15 assumes the infinitesimal crack growth. θ is assumed to be a function of applied loading and material basic properties (e.g., intrinsic threshold stress intensity factor ΔK_{th} and the fracture toughness value K_c), which is 90° when the applied loading is approaching threshold and is 0° when the applied loading is approaching the fracture toughness. Mathematically, θ can be expressed as

$$\theta = \frac{\pi}{2} - \frac{\pi}{2} \frac{\Delta K - \Delta K_{th}}{K_c - \Delta K_{th}} \tag{16}$$

According to Figure 10, when the crack tip propagate from point O to O' , the CTOD grows from δ_0 to δ'. The CTOD at the two time instants t and t + dt can be expressed as

$$\delta' = \lambda(\sigma + d\sigma)^2 (a + da) \tag{17}$$

$$\delta_0 = \lambda \sigma^2 a \tag{18}$$

λ is redefined as $\lambda = \frac{3\pi}{8E\sigma_y}$ as to include the hardening effect. Ignoring the high order terms, the change of CTOD can be expressed as

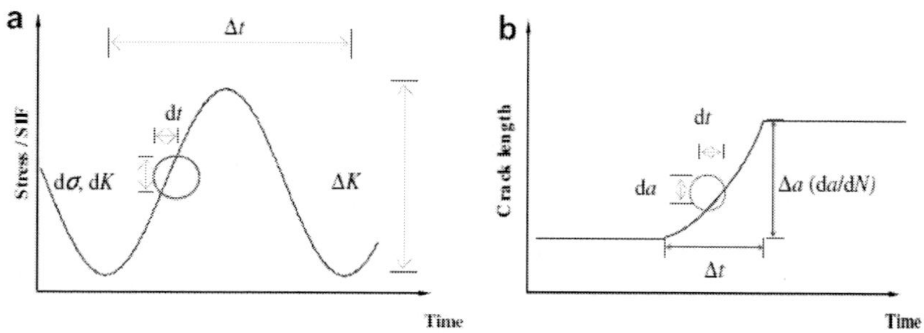

Figure 9. Schematic representation growth model (a) stress/Stress Intensity Factor (SIF) of crack tip geometry vs. Time; (b) Crack growth vs. time (Lu and Liu, 2010).

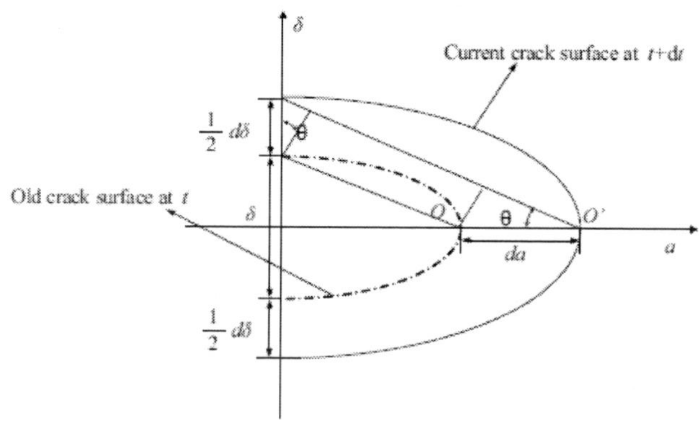

Figure 10. Schematic illustration of the small time scale crack (Lu and Liu, 2010).

$$d\delta = \delta' - \delta_0 = \lambda\left(2\sigma a d\sigma + \sigma^2 da\right) \quad (19)$$

Substituting Eq.19 into Eq.15, the instantaneous crack growth rate can be calculated as

$$\frac{1}{C\lambda a}\frac{da}{dt} = \frac{2\sigma}{1 - C\lambda\sigma^2}\frac{d\sigma}{dt} \quad (20)$$

Eq. (20) is the proposed small scale fatigue crack growth model based on the instantaneous crack kinetics (da/dt) instead of the reversal-based crack growth rate (da/dN). The crack length at any arbitrary time can be calculated via direct time integration as

$$\int_t^{t+\Delta t} \frac{1}{C\lambda a}\frac{da}{dt} = \int_t^{t+\Delta t} \frac{2\sigma}{1-C\lambda\sigma^2}\frac{d\sigma}{dt} \tag{21}$$

The relationship between the classical fatigue crack growth analysis and the proposed incremental crack growth model are discussed in this section. It is valuable to show the predictive results of the proposed incremental crack growth model applied to constant amplitude loading tests and to compare it with traditionally used reversal-based approach.

The classical da/dN curves can be reproduced by the direct integration over the time duration of one cycle, i.e., Δt. The derivation is shown below. The proposed incremental crack growth model

Eq. 20 can be rewritten as

$$\frac{da}{C\lambda a} = \frac{2\sigma d\sigma}{1-C\lambda\sigma^2} \tag{22}$$

During the loading progress, i.e. the load increases from σ_{min} to σ_{max}, integrate both sides of Eq. 22.

$$\int_{a_0}^{a_0+\Delta} \frac{da}{C\lambda a} = \int_{\sigma_{min}}^{\sigma_{max}} \frac{2\sigma d\sigma}{1-C\lambda\sigma^2} = \int_{\sigma_{min}}^{\sigma_{ref}} \frac{2\sigma d\sigma}{1-C\lambda\sigma^2} + \int_{\sigma_{ref}}^{\sigma_{max}} \frac{2\sigma d\sigma}{1-C\lambda\sigma^2} \tag{23}$$

Because crack only grows during the portion of the load cycle (i.e. when $\sigma > \sigma_{ref}$), the first term in the right hand of Eq. 23 equals to zero. The crack growth Δa in one cycle can be expressed as

$$\Delta a = \frac{da}{dN} = \frac{\pi C\lambda a\left(\sigma_{max}^2 - \sigma_{ref}^2\right)}{\pi\left(1-C\lambda\sigma_{max}^2\right)} = \frac{C\lambda\left(K_{max}^2 - K_{ref}^2\right)}{\pi\left(1-C\lambda\sigma_{max}^2\right)} \tag{24}$$

where $K_{ref} = \sigma_{ref}\sqrt{\pi a}$.

It can be seen that the predicted crack growth rate da/dN curves already includes the stress ratio effects. Eq. 24 is used for model prediction under constant amplitude loading and is validated using extensive experimental data as shown in Figure 11.

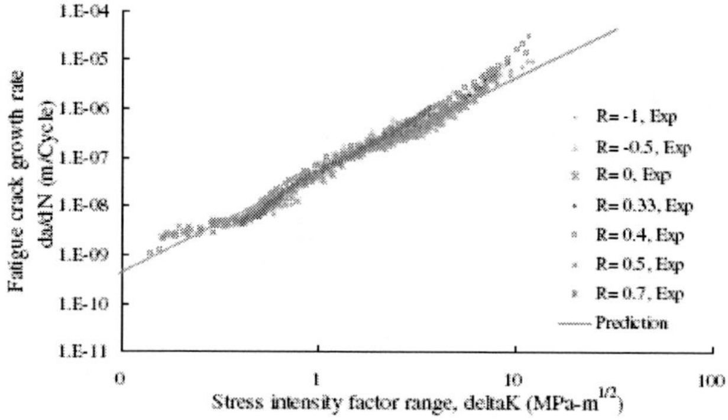

(a) Al 2024-T3

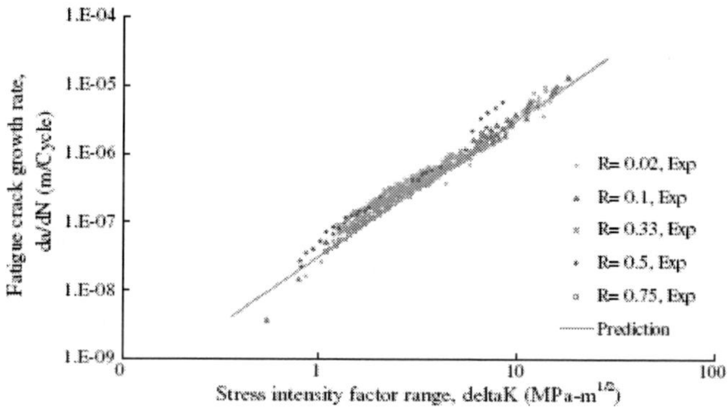

(b) Al 7050-T7451

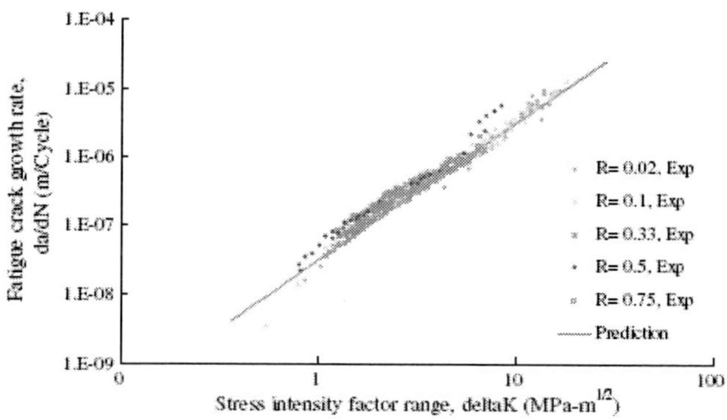

(c) Al 7075-T6

Figure 11. (Continued).

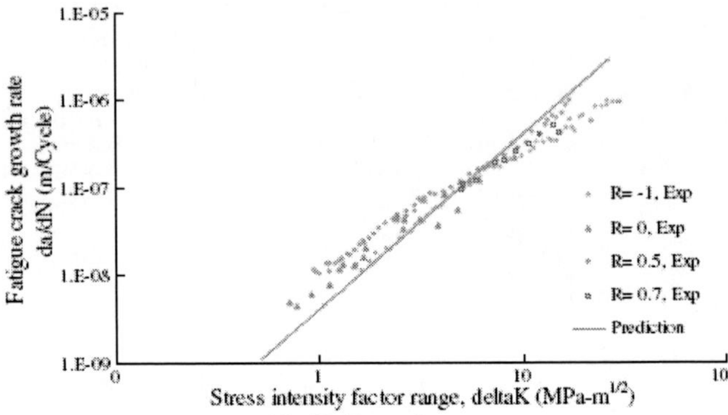

(d) 4340 steel

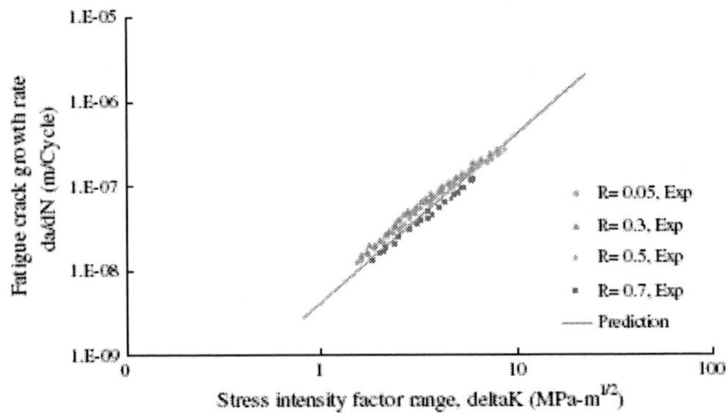

(e) 300M steel

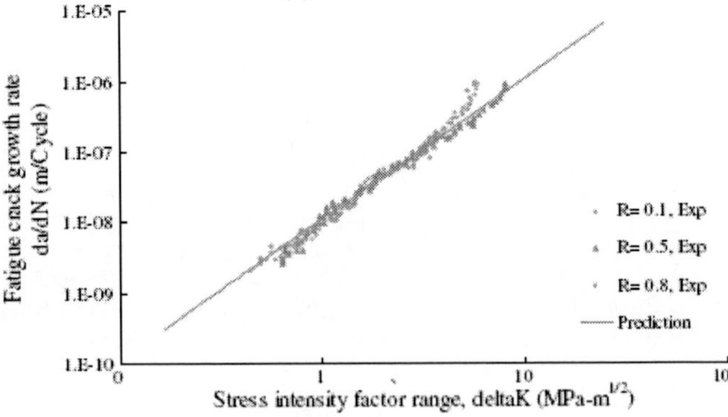

(f) Ti-10V-2Fe-3Al

Figure 11. Experimental data and model prediction using small scale crack growth model in time domain (Lu and Liu, 2010).

3. Small Crack Growth Rate Model

The behavior of small cracks growth becomes a topic of interest in recent years. Studies on the small cracks show that many materials, such as aluminum alloy and titanium alloy, have the abnormal growth phenomenon of small cracks. The common conclusion is obtained that small cracks grow faster than long cracks at the same nominal driving force and even grow at stress intensity factor ranges below the long cracks threshold. The prediction method based on the data from long fatigue cracks tests will overestimate the fatigue lives of structures with short cracks. The behavior of short fatigue cracks is a matter of importance not only because much of the fatigue lifetime is spent in propagating these cracks, but also because the boundary between propagation and non-propagation separates the safe from the potentially unsafe fatigue regimes. Small fatigue cracks differ from large cracks in three important aspects (McEvily, 2005). These are:

1. Crack growth is often elastic-plastic in nature rather than linear-elastic because of a high ratio of the fatigue strength to the yield strength and the consequent large ratio of the plastic zone size to the crack length.
2. Crack closure is a function of the crack length. In the wake of a crack of some microns in length the crack closure level is zero, but as the crack grows to a length of a millimeter or so the crack closure level rises to that of a large crack.
3. The threshold condition for fatigue crack growth is determined by the range of the fatigue strength, not the range of the long-crack fatigue crack growth threshold stress intensity factor.

3.1. Physically Small Crack Growth Model

It is well agreed that for physically small crack growth, the elastic plastic fracture mechanics (EPFM) must be employed. The EPFM approach was first introduced by Tomkins in 1968 (Miller 1999) who equated da/dN to crack tip decohesion (from knowledge of the cyclic stress-strain curve) and thence to the bulk plastic strain field as occurs, for example, under high strain fatigue, thus:

$$\frac{da}{dN} = B\left(\Delta\varepsilon_p\sqrt{\pi a}\right)^m - \Psi \qquad (25)$$

where Ψ is the threshold condition in this crack regime.

Kitagawa and Takahashi (1976) pointed out that the rate of short cracks growth is determined by the range of cyclic stress rather than the range of the stress intensity factor that controls the long fatigue cracks growth. The effect of crack size on fatigue crack growth threshold can be described by means of the Kitagawa and Takahashi plot relating the threshold stress with the crack size, as shown in Figure 12.

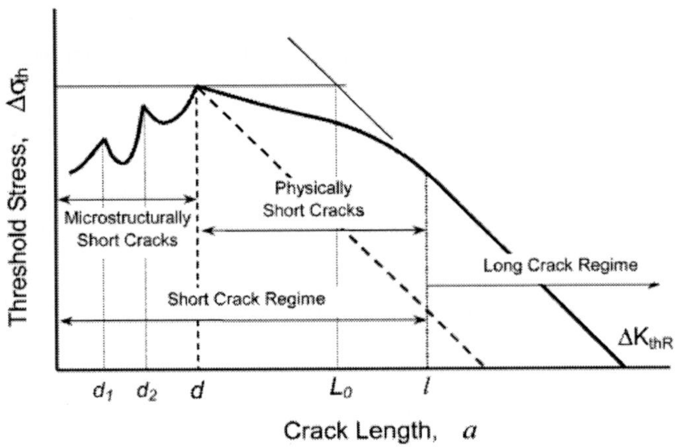

Figure 12. Kitagawa-Takahashi type diagram (1976).

Others established the relation between the rate of fatigue crack growth, da/dN, and some functions of the range of the stress intensity factor, ΔK by modifying the crack growth relation for long cracks. McEvily et al. (1999; McEvily and Ishihara 2001) proposed the following modified constitutive relationship,

$$\frac{da}{dN} = A\left\{\left[\sqrt{\pi\gamma_e\left(Sec\frac{\pi}{2}\cdot\frac{\sigma_{\max}}{\sigma_y}+1\right)}+Y\sqrt{\frac{\pi}{2}a\left(Sec\frac{\pi}{2}\frac{\sigma_{\max}}{\sigma_y}+1\right)}\right]\Delta\sigma - \left(1-e^{-ka}\right)\left(K_{op\max}-K_{\min}\right)-\Delta K_{efth}\right\}^2 \tag{26}$$

where

$$\gamma_e = \left(\frac{\Delta K_{effth}}{\Delta \sigma_{EL}}\right)^2 \cdot \frac{1}{\pi\left(Sec\frac{\pi}{2}\cdot\frac{\sigma_{max}}{\sigma_y}+1\right)} \tag{27}$$

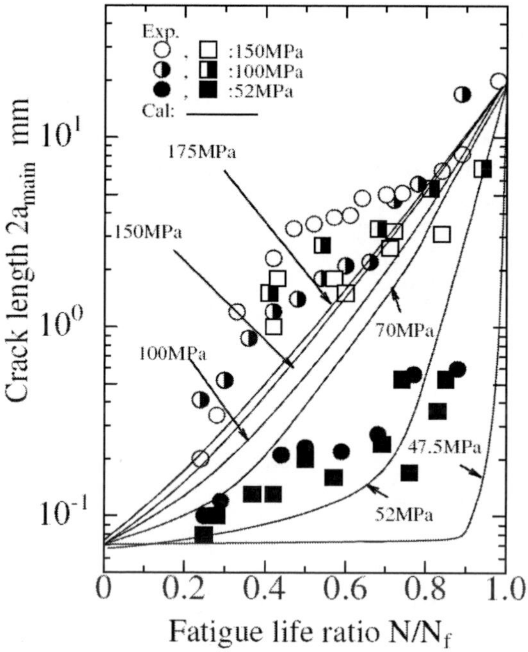

Figure 13. Experimental and calculated values of the crack length as a function of the ratio N/Nf for the aluminum alloy 2017-T4 at several stress amplitudes (Ishihara and McEvily, 1999; McEvily, 2005).

A is a material- and environmentally sensitive constant, $(MPa)^{-2}$; ΔK_{effth} is the effective range of the stress intensity factor at the threshold level, $MPam^{0.5}$; $K_{opmax}$is the maximum stress intensity factor at the opening level for a macroscopic crack, $MPam^{0.5}$; K_{min} is the minimum stress intensity factor applied, $MPam^{0.5}$; k is a material constant which reflects the rate of crack closure development with crack advance; Y is a geometrical factor; a is the actual crack length, m; σ_y is the yield strength of the material, MPa; $\Delta\sigma$ is the stress range applied, MPa; σ_{max} is the maximum stress applied, MPa; $\Delta\sigma_{EL}$ is the stress range of the endurance limit, MPa. Comparison with some experimental data showed that this constitutive relationship is able to explain the six phenomena concerned. The model can be validate by Figure 13 by comparison of experimental and calculated values of the crack length as a function of the ratio N/N_f for the aluminum alloy 2017-T4 at several stress amplitudes.

3.2. Microstructurally Small Crack Growth Model

Microstructural Fracture Mechanics (MFM) was proposed to handle the crack propagation in micro crack level. The MFM approach was first introduced by Hobson et al. (1986) and followed by Navarro and de los Rios (1988). The crack growth law is expressed as

$$\frac{da}{dN} = C\Delta\gamma^{\beta}(d-a) \tag{28}$$

where d is a microstructural dimension. It should be noted that Eq.28 indicates a zero crack speed when the crack depth, a, is equal to d, and that prior to this state the crack will continuously decelerate until it either stops or continues to propagate according to an overlapping continuum mechanics description.

4. Crack Growth Rate Model with Creep Interaction

Engineering components and structures, designed for high-temperature service (such as those in chemical plants and stationary and aircraft turbine engines), are subjected to cyclic loads with dwell time, at maximum load, of different duration that can exacerbate the development of fatigue damage through creep and environment enhanced crack growth (Wei and Huang, 2002).

The fatigue failure with creep interaction has attract attention especially in high temperature condition. Typically, linear accumulation model is used to consider the common interaction of fatigue and creep. In recent years, creep effect on fatigue in room temperature is also highlighted, which is more easily neglected in the past.

In the past four decades, significant effort and progress have been made to study dwell fatigue and pure fatigue behavior of metal alloys to find out how the material responds to dwell time. The prior research attributes dwell sensitivity to many deleterious mechanisms.

These mechanisms include: time-dependent strain accumulation, microstructural and micro-texture influences, stress ratio effects, internal hydrogen embrittlement and environmental effects, crystallographic orientation dependence, interactions between creep and fatigue. However, there is no consensus on the basic cause of the dwell fatigue sensitivity of titanium alloys. Up to now, the problem remains opened as well on the scientific as on the engineer point of view.

In 1980, Munz and Bachmann proposed the separation idea to solve the problem. Figure 14 shows the separation of cycles with hold times in cycles with a triangular wave form and a constant load. If such a separation is possible, the crack growth rate can be calculated by the addition of two terms: the crack growth rate $(da/dN)_{tiang}$ of a test with a triangular wave form an the crack growth rate daidt in a constant load test at the stress intensity factor K_{max}, corresponding to the maximum load, times the hold time Δt,

$$\frac{da}{dN}=\left(\frac{da}{dN}\right)_{triang}+\left(\frac{da}{dt}\right)_{K=K_{max}}\cdot\Delta t \tag{29}$$

Experimental studies were conducted to examine if crack growth involving hold time cycles can be adequately described by Eq. 29. Figure 15 shows da/dN versus ΔK curves for Ti-6Al-4V in air. Tests with 3Hz and 0.3Hz were performed for the triangular wave form. There is only a small effect of frequency on crack growth rate. A rapid increase in crack growth rate occurs at about $\Delta K = K_{Ith}$ (1-R) for the cycles with hold time. For 5 min hold time da/dN is larger than that for 1 min hold time.

Figure 16 is the comparison of measured and predicted crack growth rate of Ti-6 Al-4 V in air. From the measured da/dt-ΔK curve in air and the da/dt-ΔK curve for the triangular wave form the crack growth rate da/dN for the hold time cycle calculated according to Eq.29, it can be seen that for the tests in air the agreement between calculated and measured crack growth rates is good. Whereas other tests in salt solution were made which shows that much higher crack growth rates are predicted than were measured. Therefore it is only possible to predict the crack growth rate for hold time cycles in air environment but not in salt solution using Eq.29.

The linear summation in Eq.29 is a simplified crack growth evaluation methods which has been applicable to some structural integrity assessment procedure but has not considered the interaction between fatigue and creep damages.

Wakai (2003) conducted a benchmark study in predicting creep-fatigue crack growth in 316L(N) cracked plates subjected to a cyclic bending moment. The predictions according to different procedures using linear summation expression were compared with each other. Differences are found between the predictions of the methods. Comparing to the experimental data, it is found that the simplified methods exhibit conservatisms which are significantly reduced when integrating the creep curve continuously without initialisation during the experiment.

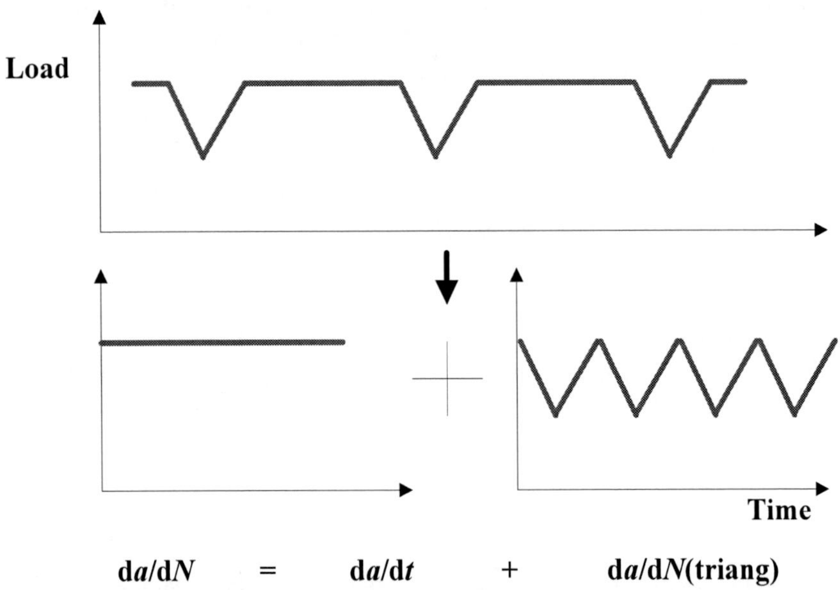

Figure 14. Separation of load cycle with hold time (Munz and Bachmann,1980).

Based on the short fatigue crack growth rate model modified according to Eq.26, a dwell-fatigue crack growth rate model is proposed by Wang et al. (2014).

Two factors are included in the dwell-fatigue crack growth rate model:(1) the effect of the dwell time at the maximum stress ; (2) the effect of the maximum stress (the dwell-fatigue life is smaller than pure fatigue life for some titanium alloys when the maximum stress σ_{max} is larger than $0.9\sigma_y$. The dwell-fatigue life is similar with pure fatigue life when the maximum stress σ_{max} is lower than $0.9\sigma_y$.). The dwell-fatigue crack growth rate model can be expressed as follow,

$$\frac{da}{dN} =$$

$$A_1\left\{\frac{\Delta K-(\Delta K_{thR}-\Delta K_{th-s})\left[1-e^{-k(a-d)}\right]-\Delta K_{th-s}}{1-\left(\frac{K_{max}}{K_{cf}}\right)^{n_1}}\right\}^{m_1} +$$

$$A_2 t\left[\frac{(\sigma_{max}-\sigma_R)}{(\sigma_y-\sigma_R)}\right]^{\lambda}\left[\frac{\Delta K}{1-\left(\frac{K_{max}}{K_{cf}}\right)^{n_2}}\right]^{m_2} \quad (16)$$

where, A_1, m_1, n_1, A_2, m_2 and n_2 are the material and environmental parameters; λ is material parameter for the influence level of the maximum stress ; t is the dwell time; d is the intrinsic crack length; $\Delta K_{th\text{-}s}$ is the crack growth threshold for short fatigue crack under stress ratio. $\Delta K_{th\text{-}cl}$ is the components of the stress intensity factor range corresponding to crack closure; Other parameters are mentioned above.

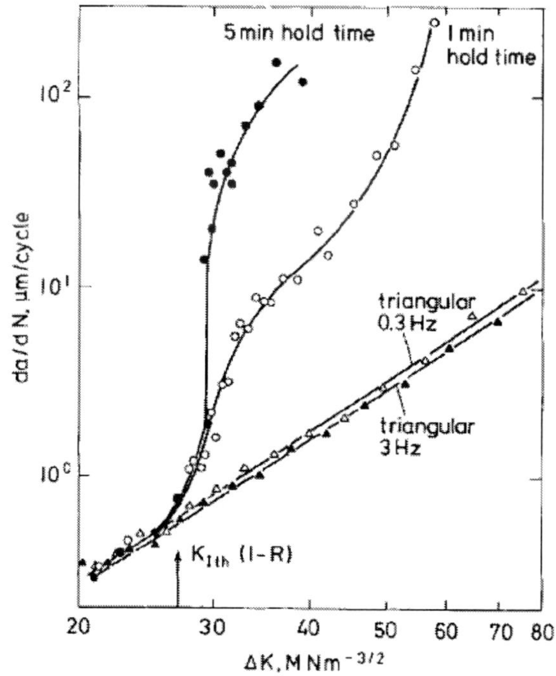

Figure 15. da/dN versus ΔK curves for Ti-6Al-4V in air (Munz and Bachmann,1980).

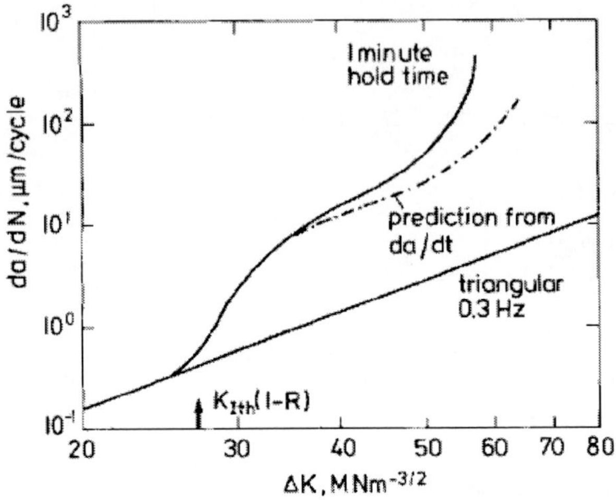

Figure 16. Comparison of measured and predicted crack growth rate of Ti-6Al-4V in air (Munz and Bachmann,1980).

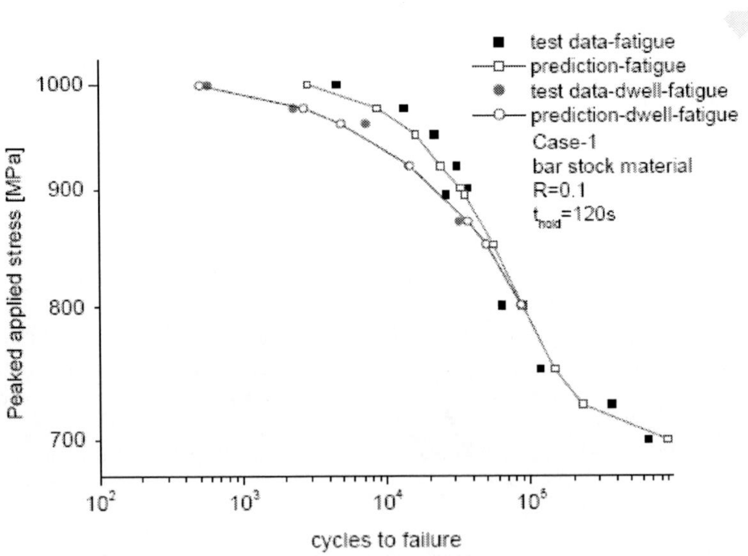

Figure 17. Comparison between the predicted curves and the experimental data of pure fatigue and dwell-fatigue lives of titanium alloy IMI834 for bimodal microstructure (Wang et al., 2014).

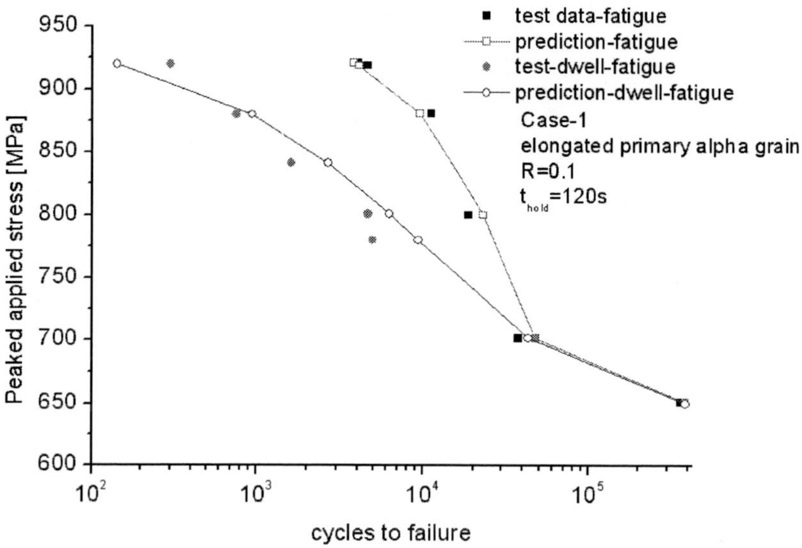

Figure 18. Comparison between the predicted curves and the experimental data45 of pure fatigue and dwell-fatigue lives of titanium alloy IMI834 for elongated primary alpha grain microstructure (Wang et al. 2014).

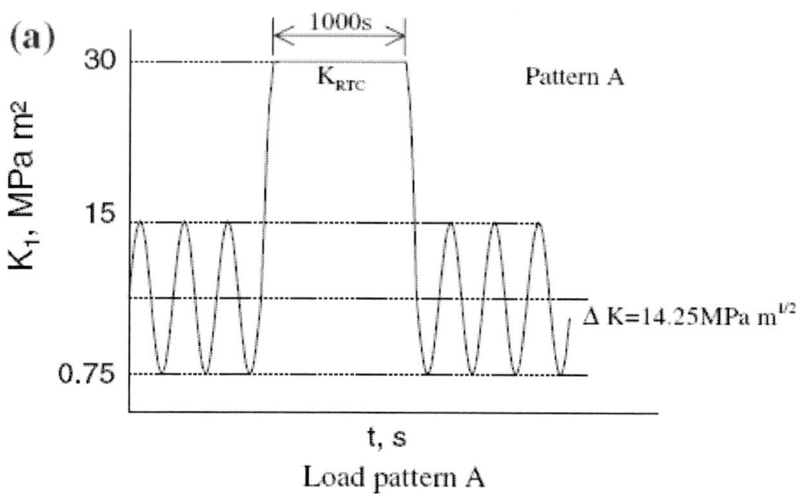

Figure 19. (Continued).

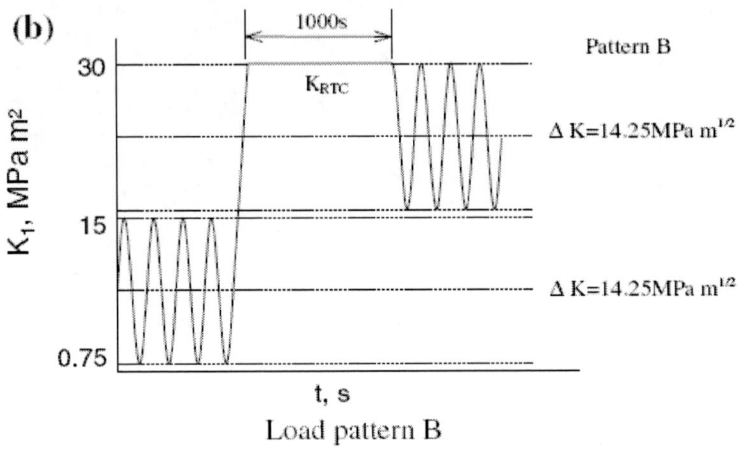

Figure 19. A schematic of load patterns of fatigue with dwell time (Zhao et al., 2006).

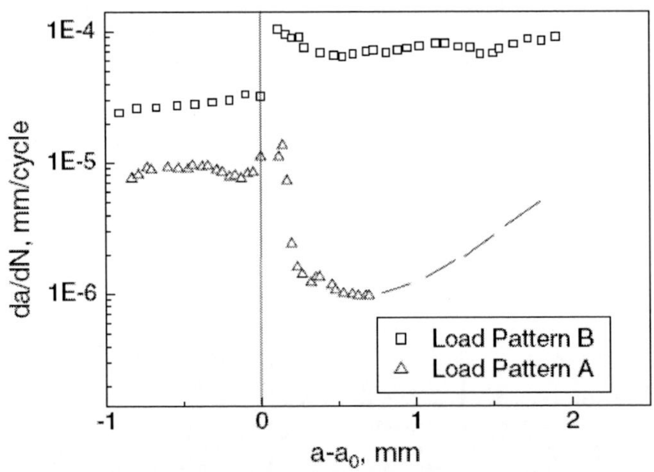

Figure 20. Influence of room temperature creep on the fatigue crack growth behavior under two load pattern (Zhao et al., 2006).

The modified model are used to predict the pure fatigue, dwell-fatigue crack growth rates and the total fatigue life of three near α titanium alloys and one α+β titanium alloy under constant-amplitude loading, and compare with experimental results. Some prediction results are compared with the experimental data as shown in Figure, 19-20.

Overload effect exists when holding stress is high than base maximum stress. Room temperature creep at a crack tip and its influence on the fatigue crack growth behavior of a 304 stainless steel have been studied by Zhao et al. (2006). The load patterns of fatigue with dwell time shown in Figure 17 are considered in the test. A time-dependent deformation has been observed at the crack tips under various stress intensity factors. The deformation increases with increasing stress intensity factor. Either acceleration or retardation of fatigue crack growth rate is found after holding time, which depends on the load pattern. Ademarcation line is observed on the fracture surface following the holding period. This implies that the crack propagation root or mode changed after the hold time. Improvement of the prediction equations should be proposed to explain more phenomenon of creep effect on fatigue crack growth rate.

REFERENCES

Bao, H., & McEvily, A. J. (1998). On plane stress–plane strain interactions in fatigue crack growth. *International journal of fatigue,* 20(6), 441-448.

Bukkapatnam, S. T., & Sadananda, K. (2005). A genetic algorithm for unified approach-based predictive modeling of fatigue crack growth. *International journal of fatigue,* 27(10), 1354-1359.

Chen, F., Wang, F., & Cui, W. (2012). Fatigue life prediction of engineering structures subjected to variable amplitude loading using the improved crack growth rate model. *Fatigue & Fracture of Engineering Materials & Structures,* 35(3), 278-290.

Cui, W. (2002). A state-of-the-art review on fatigue life prediction methods for metal structures. *Journal of marine science and technology*, 7(1), 43-56.

Cui, W. C., & Huang, X. P. (2009). A general constitutive relation for fatigue crack growth analysis of metal structures. *Acta Metallurgica Sinica (English Letters),* 16(5), 342-354.

Cui, W., Wang, F., & Huang, X. (2011). A unified fatigue life prediction method for marine structures. *Marine Structures,* 24(2), 153-181.

Donahue, R.J., Clark, H.M., Atanmo, P., Kumble, R. and McEvily, A.J. (1972), Crack opening displacement and the rate of fatigue crack growth, *International Journal of Fracture Mechanics,* Vol.8, pp.209-219.

Foreman, R.G., Kearney, V.E. and Engle, R.M. (1967), Numerical analysis of crack propagation in cyclic-loaded structures, *Journal of Basic Engineering,* Vol.89, pp.459-464.

Forsyth, P.J.E. (1969), The physical basis of metal fatigue, American Elsevier Publishing CO. New York.

Hobson, P.D., Brown, M.W. and de los Rios, E.R. (1986), EGF (ESIS) Publ. No.1, *"The Behaviour of Short Fatigue Cracks"* (K.J. Miller & E R de los Rios, Eds.), Mech. Engng, Publications Ltd., London, pp.441-459.

Ishihara, S., & McEvily, A. J. (2002). Analysis of short fatigue crack growth in cast aluminum alloys. *International Journal of Fatigue,* 24(11), 1169-1174.

Ishihara, S., McEvily, A. J., Goshima, T., Nishino, S., & Sato, M. (2008). The effect of R value on the number of delay cycles following an overload. *International Journal of Fatigue,* 30(10), 1737-1742.

Kitagawa, H., & Takahashi, S. (1976). Applicability of fracture mechanics to very small cracks or the cracks in the early stage. In *Second International Conference on Mechanical Behavior of Materials.* ASM, Metals Park, Ohio. 1976, 627-631.

Kujawski, D. (2001). Enhanced model of partial crack closure for correlation of R-ratio effects in aluminum alloys. International Journal of Fatigue, 23(2), 95-102.

Kujawski, D., & Ellyin, F. (1995). A unified approach to mean stress effect on fatigue threshold conditions. *International Journal of Fatigue,* 17(2), 101-106.

Lu, Z., & Liu, Y. (2010). Small time scale fatigue crack growth analysis. *International Journal of Fatigue,* 32(8), 1306-1321.

McEvily A.J., Yang Z. (1990). The nature of the two opening levels following an overload in fatigue crack growth. *Metall Trans A,* 21A, 2717–27.

McEvily, A. J., & Groeger, J. (1977). *Fracture* 1977. University of Waterloo Press, 2, 1293-1298.

McEvily, A.J. (2005), A method for the analysis of the growth of short fatigue cracks, *Materials Science Forum,* 482, 3-10.

McEvily, A.J. and Ishihara, S. (2001). On the dependence of the rate of fatigue crack growth on the sigma(n)(a)(2a) parameter. *International Journal of Fatigue* 23, 115-120.

McEvily, A.J., Bao, H. and Ishihara, S. (1999). A Modified Constitutive relation for fatigue crack growth, in *Proceedings of the seventh International Fatigue Congress (Fatigue'99)*, ed. X.R. Wu and Z.G. Wang, Beijing, China, Higher Education Press, 329-336.

Miller, K.J. (1999), A historical perspective of the important parameters of metal fatigue and problems for the next century. In *Proceedings of the seventh International Fatigue Congress (Fatigue'99)*, ed. X.R. Wu and Z.G. Wang, Beijing, China, Higher Education Press, 15-39.

Munz, D., & Bachmann, V. (1980). Effect of hold time and environment on fatigue crack growth rate in Ti alloys. *Materialwissenschaft und Werkstofftechnik,* 11(5), 168-172.

Navarro.A. and de los Rios, E.R. (1988), A microstructurally short fatigue crack growth equation, *Fatigue Fract. Engng. Mater. Struct.,* V-ol.11, No.5, pp.383-396.

McEvily, A.J. and Ishihara, S. (2001), On the dependence of the rate of fatigue crack growth on the $\sigma_a^n(2a)$ parameter, *International Journal of Fatigue,* Vol. 23, pp.115-120.

Paris, P.C., Gomez, M.P. and Anderson, W.P. (1961), A rational analytical theory of fatigue, *The trend in Engineering,* Vol.13, pp.9-14.

Ritchie, R.O. (1986), Small fatigue cracks: a statement of the problem and potential solutions, *Materials Science and Engineering,* Vol. 84, pp.11-16.

Rushton, P. A., & Taheri, F. (2003). Prediction of crack growth in 350WT steel subjected to constant amplitude with over-and under-loads using a modified wheeler approach. *Marine structures,* 16(7), 517-539.

Sadananda, K., & Vasudevan, A. K. (2003). Fatigue crack growth mechanisms in steels. *International Journal of Fatigue,* 25(9), 899-914.

Vasudeven, A. K., Sadananda, K., & Louat, N. (1994). A review of crack closure, fatigue crack threshold and related phenomena. *Materials Science and Engineering*: A, 188(1), 1-22.

Wakai, T., Poussard, C., & Drubay, B. (2003). A comparison between Japanese and French A16 defect assessment procedures for creep-fatigue crack growth. *Nuclear engineering and design*, 224(3), 245-252.

Wang, F., Chen, F. L., & Cui, W. C. (2010). Applicability of the improved crack growth rate model and its parameters estimation method. *Journal of Ship Mechanics,* 14(3), 252-262.

Wang K., Wang F., & Cui, W. C. (2014). *Prediction of dwell-fatigue crack growth of titanium alloys, Acta Metallurgica Sinica*, accpeted.

Wang, Y., Cui, W., Wu, X., Wang, F., & Huang, X. (2008). The extended McEvily model for fatigue crack growth analysis of metal structures. *International Journal of Fatigue*, 30(10), 1851-1860.

Wei, R. P., & Huang, Z. (2002). Influence of dwell time on fatigue crack growth in nickel-base superalloys. *Materials Science and Engineering:* A, 336(1), 209-214.

Zhao, J., Mo, T., Nie, D. F., Ren, M. F., Guo, X. L., & Chen, W. X. (2006). Acceleration and retardation of fatigue crack growth rate due to room temperature creep at crack tip in a 304 stainless steel. *Journal of materials science,* 41(19), 6431-6434.

In: Recent Trends in Fatigue Design
Editor: Ricardo Branco
ISBN: 978-1-61668-410-5

Chapter 6

Short Fatigue Crack Behavior in Duplex Stainless Steels

M. C. Marinelli* and S. A. N. Hereñú
Instituto de Física Rosario – Consejo Nacional de Investigaciones Científicas y Técnicas (CONICET),
Universidad Nacional de Rosario, Argentina

Abstract

The aim of this study is to analyze the short fatigue crack behavior of different duplex stainless steel (DSS) grades under two thermal conditions: as-received and embrittled (100 h at 475oC). A detailed review on the state of art of this theme has been carried out.

The chapter is divided into five sections. The first section or introduction outlines the theoretical background necessary for the present investigation. In the second section an overall description of the principal experimental techniques employed is highlighted. In the third section the results reported from different researchers are summarized and discussed. This third section is divided in two subsections: low cycle fatigue (LCF) and high cycle fatigue (HCF), due to the different physical mechanisms involved in each case to produce failure. In the fourth section a model to reproduce the propagation of microstructural fatigue cracks in DSS is adapted. In the last section the concluding remarks and scope for further research in this area of investigation are presented.

* Corresponding Author address. Email: marinelli@ifir-conicet.gov.ar.

1. INTRODUCTION

1.1. Duplex Stainless Steels

Stainless steel are iron-base alloys with a minimum chromium content of 11wt% which is responsible for developing, in the presence of oxygen, a passive film on its surface. This film protects the material from corrosion processes. By properly adjusting the chemical composition and thermal treatments of the stainless steels, four types of microstructure can be obtained, austenitic, ferritic, martensitic, duplex [1]. Duplex stainless steels (DSS) contain principally as alloying elements Cr (22 to 25%) and Ni (5 to 7%). Copper, molybdenum, tungsten, nitrogen, manganese, and silicon may be added to control phase balance and to improve corrosion resistance. The composition of the DSS and its heat treatment leads to different amount of the ferrite and austenite. Different advantages are obtained with this mixed ferrite/austenite microstructure if compared with the unique structure of austenitic and ferritic stainless steel. DSS show more toughness than most ferritic grades and double the yield strength of the common austenitic grades. Moreover, they have better stress corrosion crack (SCC) resistance than most austenitic grades. Due to lower nickel content, DSS are also less expensive than austenitic stainless steels with similar corrosion resistance. The combination of properties, including high strength and excellent corrosion resistance and chloride stress corrosion cracking (SCC) made DSS attractive for many purposes. Duplex stainless steel are mainly used in petrochemical industries, pulp and paper mill industries, chemical industries, food and equipments, offshore platforms, structural and mechanical components, sulfuric acid plants, containers for transportation [2-6].

The main evolution in the chemical composition of DSS has been the progressive increase in the nitrogen content which leads to three commercial grades: first generation with N lower than 0.1%; second generation with nitrogen between 0.1% and 0.2%; and third generation with more than 0.2%N. The addition of nitrogen as an alloying element leads to remarkable hardness, improvements in yield stress and corrosion resistance and, also reduces the rate at which detrimental intermetallic phases form. Within the second generation DSS, the 22Cr-5Ni-3Mo variant (commercially designed as SAF 2205 or UNS S32205) is considered the standard duplex type. On the other hand, superduplex stainless steels (UNS S32750) included in the third generation DSS were developed to resist corrosion in seawater. The hardness of the austenite phase increases because of the nitrogen content whereas chromium and molybdenum enhance the hardness of both phases (mostly ferrite). Recently, a new family of economical duplex stainless steels, lean duplex stainless steel (LDSS), in which N or Mn substitute the expensive Ni, has been introduced in the market. Lean duplex stainless steel can certainly improve the progress of the duplex steels since it is possible to retain quality and reduce cost [7, 8].

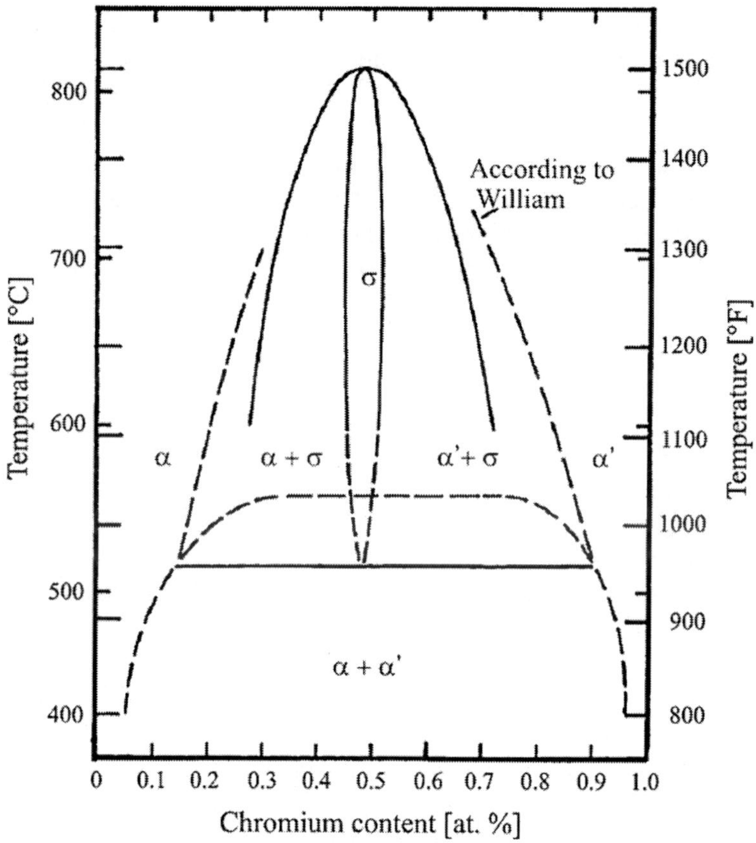

Figure 1. Phase diagram iron-chromium [9].

1.2. 475°C Embrittlement

The high chromium concentration in DSS not only enhances their excellent resistance to pitting and crevice corrosion but also makes them susceptible to embrittlement when they are exposed for long aging time in the temperature range of 300-500°C. This phenomenon is attributed to the presence of a miscibility gap in the Fe–Cr system where the homogeneous ferrite phase decomposes into α (Fe-rich) and α' (Cr-rich), Fig.1 [9]. Inside the spinodal curve the decomposition occurs through spinodal decomposition (SD), whereas immediately outside the spinodal line, but still within the miscibility gap, decomposition may take place by classical nucleation and growth. Spinodal decomposition starts with a fluctuation in Cr content, whose amplitude and wavelength grow up to the equilibrium stage. This phase separation process occurs at a very fine scale (of the order of only a few nanometers) and the presence of α' phase can only be detected through an atom probe field ion microscope. By means of transmission electron microscopy (TEM), SD can be identified diffusely as a modulated contrast called "orange peel". The presence of α', irrespective of the reaction by which it is formed, causes embrittlement and increases the hardness of the ferrite phase. The rate of embrittlement is highest at 475°C and is referred to as "475°C embrittlement". This aging effect detrimentally affects impact toughness, tensile ductility, fracture toughness and fatigue behavior, limiting the use of these steels at temperatures below 300 °C [10-12].

1.3. Fatigue Behavior

The fatigue process actually considers two domains of cyclic stressing or straining where failure occurs by apparently different physical mechanisms: one where significant plastic straining occurs and the other where stresses and strains are confined to the elastic region. The first domain, referred as low cycle fatigue (LCF), involves relatively short lives (10 to 100.000 cycles). The other domain, referred to as high cycle fatigue (HCF), is associated with long lives (greater than 100.000 cycles). LCF leads to a relatively homogenous macro-plastic deformation in all grains. The dislocation slip or twinning is observed in a large number of grains. In the HCF regime, only in a fraction of favorably oriented grains show plastic deformation. The eventual failure at a stress below the yield strength of the material in the HCF regime can be rationalized by the irreversible nature of cyclic slip. The cyclic slip irreversibility is extensively researched [14-16].

The degradation of material resulting from cyclic stress/strain is called fatigue damage. The fatigue damage for common metals and alloys consists of the following stages:

- **Substructural changes causing cyclic strain localization**

  This process encompasses the formation and stabilization of a characteristic dislocation structure associated with the initial hardening or softening. This is followed by the localization of the cyclic plastic strain into bands called persistent slip bands (PSB).

- **Short crack initiation**

  The most common features observed by scanning electron microscopy (SEM) on a polished free surface of a specimen subjected to cyclic loading are the slip markings and surface reliefs [18]. The slip markings are intersections of the active slip planes with the surface. During fatigue the presence of slip markings does not necessarily mean that a crack will initiate there [19]. Small cracks nucleate mainly at or near the surface where local plastic strain concentration is higher due to the higher stresses present there and a lower degree of constrains of the near surface volumes. In areas where lamellae of localized cyclic slip – persistent slip bands (PSBs) – intersect the original flat surface, surface relief, known as surface extrusions and intrusions, arises [20]. These surface roughness causes stress concentrations leading to fatigue crack nucleation. Moreover, local plastic concentration at the surface could also be exerted by slip bands, inclusions, grain boundaries, surface defects and twin boundaries leading the nucleation of small fatigue cracks.

- **Growth and coalescence of initiated short cracks to form a dominant macro-crack**

  Though the initiation of fatigue cracks is an extremely important phenomenon for the evolution of the fatigue damage usually represents 1 to 20% of fatigue life of a smooth body. The majority of fatigue life is spent in the growth of short cracks. Therefore, the growth of short fatigue cracks attracts such a high interest [21].

  Once a small crack is nucleated, it propagates along the persistent slip bands on planes of maximum shear stress. This frequently is called stage I crack growth. The growth rate within this stage is a nonlinear and noncontinuous. This discontinuity in small crack propagation can be attributed to microstructural barriers (such as grain boundaries, twin boundaries, interphases) that block or temporally arrest cracks until attaining a critical stress value. In a polycrystalline metal the crack may extend for only a few grain diameters before the crack propagation changes to stage II.

- **Stable propagation of the dominant macro-crack**

  In stage II crack growth involves growth of well-defined crack in a direction normal to maximum tensile stress. The rate of crack propagation in stage I is generally very low as compared to the crack propagation rates for stage II [22].When a small crack reaches a critical length, microstructural barriers stop to exert any influence on its propagation rate, i.e. crack follows a continuous and lineal law for its propagation and cracks are considered long cracks.
- **Unstable propagation of the dominant macro-crack causing structural instability and final failure**

  A substantial part of the lifetime of a component which finally fails will be while the crack is short (i.e. of the order of a few grains in size). Under these conditions, its growth may be very sensitive to the local microstructure. In DSS there are several factors such as orientation relationship between the phases, partitioning of alloying elements to different phases, texture, and topography can affect short crack behavior. Furthermore, when one of the phases undergoes aging embrittlement a new variable affects the fatigue behavior of the material.

2. MATERIALS AND EXPERIMENTAL PROCEDURES

The DSS mainly used in this work are: standard UNS S32205, superduplex UNS S32750 and lean UNS S32003. Table 1 gives the chemical composition in weight percent of all the DSS studied. The as-received DSS UNS S32205 and UNS S32750 were manufactured in the form of hot-rolled and solution annealed cylindrical bars. DSS UNS S32750 was also supplied in the form of 3mm plate. The as-received LDSS were supplied in longitudinally welded stainless steel pipes.

Table 1. Chemical composition in wt% of the DSS

	C	Ni	Cr	Mo	N	Mn	P	Si	Cu	S	Fe
S32205	0.03	5	22	3.2	0.18	1.2	0.03	1		0.015	Balance
S32750	0.015	7	24.5	3.8	0.26	0.4	0.015	0.3	0.31	0.0005	Balance
S32003	0.021	3.8	22	1.8	0.18	1.73	0.024	0.22			Balance

Microhardness measurements were carried out in both phases of each DSS. The Vickers indentations were performed with a load of 245.2 mN during 10 s. For each sample a minimum of 10 indentations were performed, Table 2.

Table 2. Vickers hardness of the DSS as-received

	Ferrite	Austenite
S32205 AR	340 ± 20	330 ± 20
S32750 AR	370 ± 20	340 ± 20
S32003 AR	310 ± 10	320 ± 20

The grain sizes of the as received DSS were very small for obtaining discernible crystallographic data from interrupted fatigue tests. Therefore, the grains were sometimes coarsened through the following heat treatment: homogenization for 4 h at 1250°C; slow cooling to 1050°C and finally water quenching. The microstructure consists of a balanced proportion of austenite in ferrite.

It is assumed that the spinodal decomposition in the ferritic phase is already completed at 475 °C during 100 h. Thus, in order to analyze the effect of 475 °C embrittlement on the fatigue behavior of DSS, some specimens were aged for 100 h at 475 °C. These specimens are referred as aged.

Before testing, all the specimens were initially ground and polished with sequentially finer grits. Finally, the specimens were additionally electrolytic polished to facilitate observations during LCF tests.

LCF tests were conducted at room temperature under fully reversed plastic strain or total strain control. In order to observe the fatigue specimen surface damage during cycling, the LCF tests were stopped systematically and a selected area was photographed using a high resolution CCD camera mounted on the fatigue test machine. The images were stored to record the initiation, growth and propagation of the microcracks. HCF tests were carried out in a servo-hydraulic testing system under stress control.

After the different mechanical tests, the examination of the microcracks was carried out with scanning electron microscopy (SEM) in secondary and backscattered electron contrast modes. Electron back scattered diffraction (EBSD) technique and orientation imaging microscopy (OIM) were used to relate the crystallographic features of the sample surface with the evolution of surface damage during fatigue testing. Internal dislocation structures were studied by transmission electron microscopy (TEM) operated at 100 kV.

3. EXPERIMENTAL RESULTS

3.1. LCF

3.1.1. DSS As-received

Detailed analysis of optical images and strain field measurements by digital image correlation were reported in X2 Cr Ni Mo 25-07 nitrogen alloyed DSS [23]. The microstructure of this DSS is 40% vol. austenite and 60% vol. ferrite. The austenitic grains are equiaxial with a mean grain size of 10 μm. The ferritic grains are approximately 50 μm length and 10 μm of width. The cyclic test were performed at the total strain amplitude of $\Delta\varepsilon_t = 1 \times 10^{-2}$. This research assumes that initially the austenitic phase accommodates the cyclic plastic strain due to strain heterogeneities firstly observed in this phase. Slip markings appears latter in the ferritic phase. Although the intensities of the slip markings in the austenitic phase were stronger than those observed in some ferritic grain, the first microcracks initiated in the ferritic phase, preferentially at ferrite/ferrite grain boundaries. In the austenite, the nitrogen alloying changes the slip mode from wavy to planar and increases the reversibility of dislocation slip. Thus, microcracks nucleation in ferrite can be rationalized by the irreversibility of the dislocation slip due to easy cross slip that leads to pronounced extrusions and intrusions where the crack nucleates. Atomic force microscopy studies of the surface relief evolution in the two phases of a DSS 25Cr-7Ni-0,25N subjected to cyclic deformation [24] agrees with El Bartali et al. proposal [23]. Inside the ferrite grains different types of extrusions were found: rectilinear band-like extrusions, curvilinear cord-like extrusions with height of 15 to 50nm and highly rugged areas with height of approximately 100nm, HR, situated mainly along phase boundaries. These HR areas concentrate the deformation and lead to crack nucleation. These authors assume that HR areas are due to plastic irreversibility of BCC caused by the slip difference of screw dislocations in traction and compression. On the other hand, Balbi et al. [25] correlate the fatigue damage with the internal dislocation in the as-received DSS UNS S32205 tested at $\Delta\varepsilon_p = 3x10^{-3}$. They propose that the generation of PSBs in the ferrite phase controls the nucleation and growth of a microcrack. This mechanism provides sites for microcrack nucleation at grain boundaries and paths for microcrack propagation.

The misorientation between neighbouring grains and the slip system difference between two phases takes great part of the strengthening of polycrystalline materials as it limits dislocation movement and causes internal stresses. This phenomenon finally leads to strain localizations at grain or phase boundaries and causes internal stresses [26]. In DSS a crystallographic orientation relationships exist between ferrite (BCC) and austenite (FCC) grains, typically [110](1-11)γ //[111](1-10)α , referred as K-S. Some easy slip directions in one phase coincide with easy slip directions in the other phase when this K-S relationship is fulfilled and no strain localization is seen. Nevertheless, some phase boundaries do not exhibit this phenomenon and a continuous and linear slip markings are not observed through $\alpha/\ \gamma$ boundaries. Instead shear bands arrested at the phase boundary frequently causes crack nucleation [27]. Some researches [28-30] reported the important role of crystallographic correlation between phases in DSS. Marinelli et al [28] analyzed the fatigue damage of DSS UNS S32750, which contains approximately equal volume fractions of austenite and ferrite as well a homogeneous grain structure with a grain size of approximately 10 μm in both phases, at different deformation strain. At $\Delta\varepsilon_t = 0.8 \times 10^{-2}$ ($\Delta\varepsilon_p = 2.5\text{x}10^{-3}$), the first slip lines appeared in the austenitic phase (10 cycles). As cycling proceeds, slip lines intensify and turn into coarse bands and finally propagate into the neighbouring grains (α phase) as shown in, Figure 2a). Therefore, the deformation is transferred from one phase to the other. Finally, it was observed that microcracks nucleate mainly in the bands of the austenitic grains, Figure 2b). In order to connect intense slip marks with the near-surface dislocation structure, observations by TEM of samples just beneath the surface reveal that when K-S relationships are satisfied, the slip bands easily cross phase boundaries, Figure 2c). On the contrary, when the correlation between phases is lacking, microcracks initiated at the slip band gets arrested at the phase boundary, Figure 2d).

Finally, these authors observed at higher strain amplitude, $\Delta\varepsilon_p = 6.0\text{x}10^{-3}$, no role of K-S relations. Both phases exhibit an almost homogenous distribution of plastic deformation, however extrusions/intrusions and microcracks have been observed preferentially in the ferrite. At this strain level, the ferrite accommodates the plastic deformation through the development of a well-formed cell dislocation structure, Figure 3a). Moreover, the formation of pronounced extrusions in the ferrite phase leads to stress concentrations resulting in the formation of localized microcracks, Figure 3b). Indeed, this work stated that this higher plastic activity in the surface areas at this strain range level can be explained by the loss of the mobile edge dislocations emerging at the free surface and the corresponding accumulation of less mobile screw dislocations in this area [31]. In this process of losing edge dislocations, a rough surface is developing, which turns into extrusions and intrusions with the number of cycles.

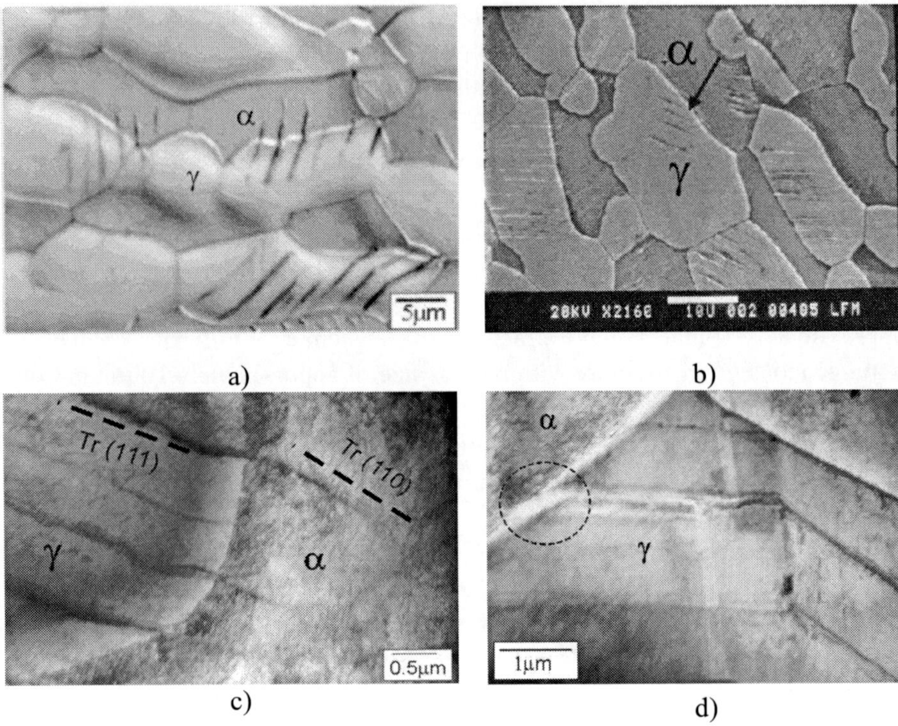

Figure 2. Surface thin foil $\Delta\varepsilon_p = 2.5\text{x}10^{-3}$, Optical micrograph a) showing intrusions/extrusions in austenitic grain and crossing phase boundary b) Short crack nucleation in austenitic grain. Surface thin foil c) dislocation bands crossing a phase boundary, d) short crack initiation in γ /α grain boundary where K-S is not satisfied [28].

The existence of the K-S relationship in flat specimens of hot- and cold-rolled plates DSS UNS S32750, cycled tested at $\Delta\varepsilon_p = 0.15\%$, was analyzed in [29]. The volume fractions of austenite and ferrite in the present DSS are approximately 54% and 46%, respectively. Both phases are polycrystalline with mean grain sizes evaluated to 10 μm in the ferrite phase and 5μm in the austenite measured perpendicular to the rolling axis. The crystallographic K-S relationships were seldom satisfied in this work. At the end of fatigue life few austenitic grains show slip marking. On contrary, during cycling the ferrite grains develop intense slip markings that turn into extrusions that concentrate plastic deformation, Figure 4. Therefore, this could explain the higher plastic activity in the ferrite. Thus, the concentration of stresses in the ferrite increases showing a significant relief in this phase as in [28] but with a lower $\Delta\varepsilon_p$ imposed. This fact could be explained by the lower grain size of austenite that leads to sustain most of the plastic deformation to the ferrite. As cycling proceeds, slip markings intensify in ferrite and turn into coarse bands that become arrested at the grain boundary. Consequently, microcracks nucleate either in extrusions or at grain boundary. Thus, the manufacturing process of cold rolling seems to modify the original crystallographic orientations between austenite and ferrite grains so the K-S relations between austenitic and ferritic grains almost disappear [32].

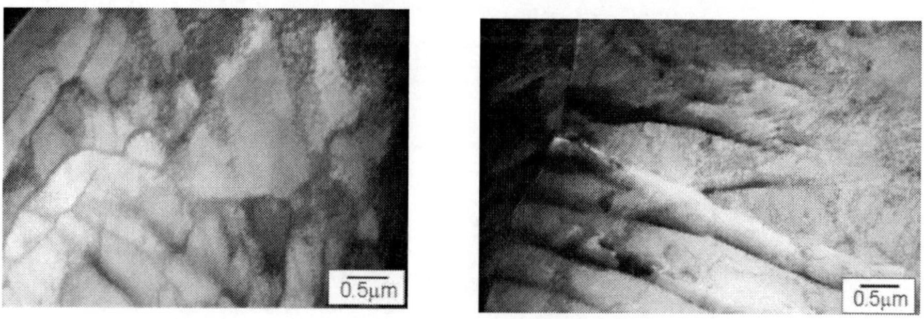

a) b)

Figure 3. Surface thin foil $\Delta\varepsilon_p$= 6 x10^{-3} a) Cell structure and incipient shear band b) cellular structure and extrusions in ferritic grain [28].

Marinelli et al [30] also studied the cyclic plastic slip activation and the short crack propagation in a cyclically loaded test at $\Delta\varepsilon_t = 1 \times 10^{-2}$ ($\Delta\varepsilon_p = 4\times10^{-3}$) of a standard DSS alloyed with 0.17% nitrogen. The volume fractions of austenite and ferrite are approximately 40% and 60% respectively. The mean grain sizes are 10μm for the austenite and 30μm for the ferrite. The K–S relations are seldom fulfilled between grains of the different phases in this case. This fact has two consequences, first, gliding dislocations cannot easily cross phase boundaries during cycling, and second, when an intense irreversible plastic activity develops in austenitic grains, several slip systems are activated in the neighboring ferritic grains, where a high relief is developed, Figure 5. Therefore, the microcracks initiate usually in α/α grain boundaries. It propagates then alternately in α/α grain boundaries, in α or γ grain, along slip markings, and only sometimes in α/γ phase boundaries. As

K–S relations are not observed, no short crack is observed propagating along 'in line' slip markings and crossing a phase boundary.

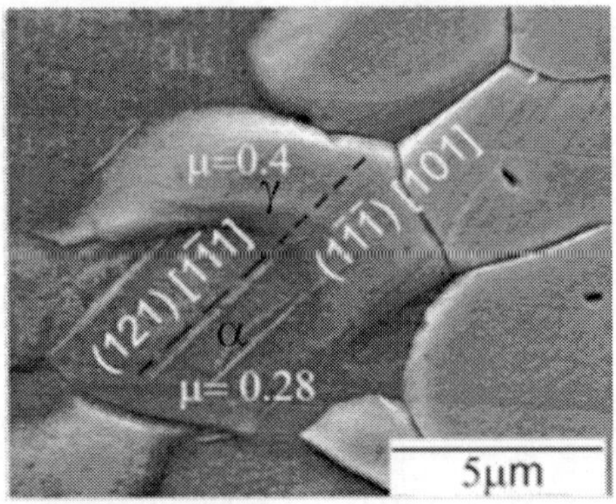

Figure 4. Higher plastic activity in ferritic grain. K-S relationship not fulfilled [29].

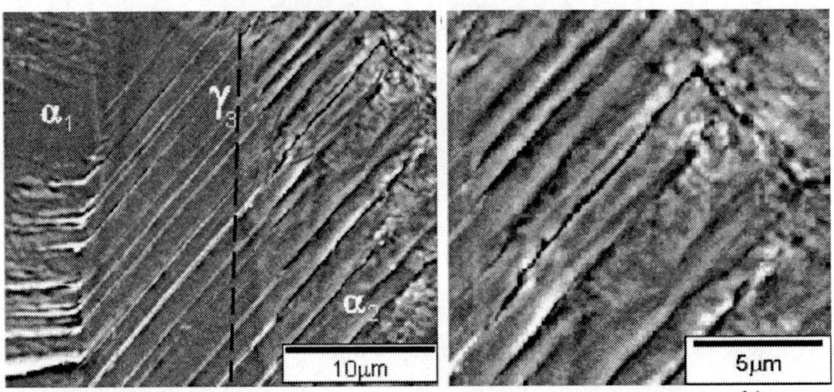

a) b)

Figure 5. SEM photographs taken at the end of the test. a) Austenitic grain γ_3 between by two ferritic grains, $\alpha_1$and α_2. Crack initiation in an intrusion in ferritic grain α_2, which does not propagate in neighboring austenitic grain, because K–S relations are not fulfilled between grains α_2 and γ_3. b) higher magnification of crack nucleation [30].

Strubbia et al [33] focused on correlating the dislocation structure with the initiation and evolution of fatigue short cracks in LDSS UNS S32003. The microstructure of these steels consists mainly of a lamellar structure of austenite islands embedded in ferrite matrix with equal volume fraction of austenite and ferrite. The average grain size is 7 ± 4 μm in ferrite and 4 ± 2μm in the austenite. Cyclic tests were conducted at room temperature under fully reversed plastic strain control, with a plastic strain range of $\Delta\varepsilon_p = 2 \times 10^{-3}$. In situ optical observations during LCF reveals that the first deformation lines appear in the ferrite phase after 100 ± 50 cycles (Figure 6a) while in the austenitic phase they appear after 300 ± 100 cycles (Figure 6b). As cycling proceeds the surface damage evolution is mainly concentrated in the ferrite (Figure 6c). Some of this slip lines in ferrite evolve to extrusions at the surface where short cracks finally nucleate, Figure 7. As the hardness of austenite phase in the present LDSS is similar than that of the ferrite, it is expected that the ferrite phase experiences most of the plastic deformation [34]. The fact agrees with the higher dislocation density developed during fatigue in the ferrite phase in comparison to the austenite phase. Hence, microcracks nucleation in ferrite during LCF can be explained by the larger plastic activity found in this phase. When the cracks reach the austenite neighbouring grain, it grows by combined slip on adjacent slip planes or along favorable slip planes. The path of a short crack is seen in Figure 8. From this figure, it is evident that phase boundaries and grain boundaries are effective barrier against crack propagation. As well as reported in standard DSSs [25], the microcracks growth in the present LDSS follows an exponential law.

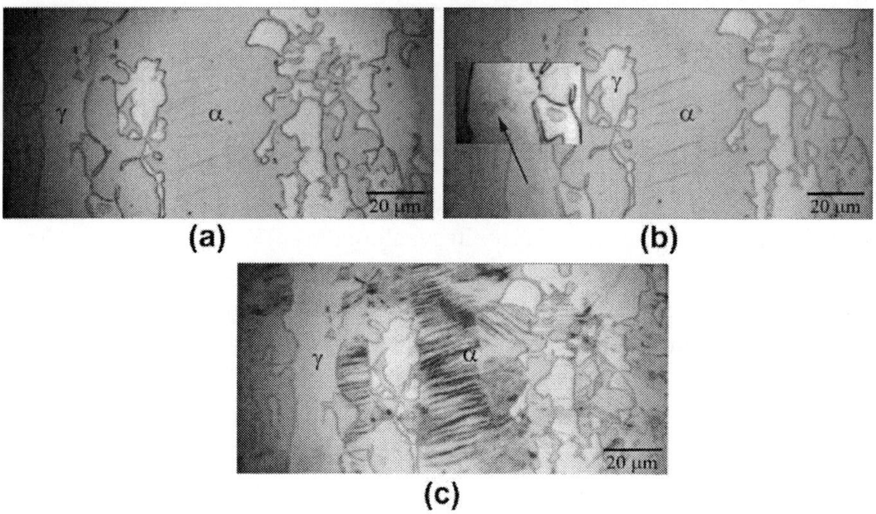

Figure 6. Slip markings at the surface of LCF specimens after: (a) 100 cycles (b) 300 cycles and (c) 15900 cycles [33].

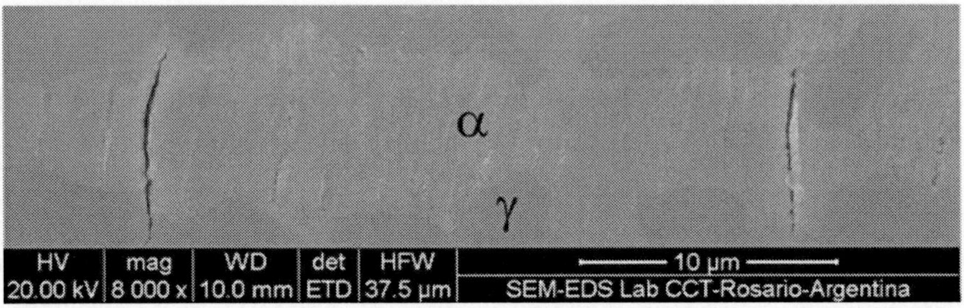

Figure 7. Fatigue damage and microcracks after 12700 cycles [33].

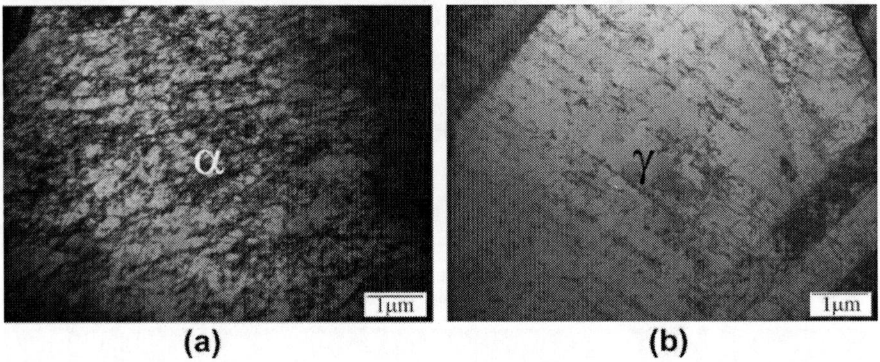

Figure.8. Dislocation structure in LCF specimens at failure (a) ferrite and (b) austenite [33].

3.1.2. DSS Aged

Aging DSS at 475°C increases the stress levels but have small effects on the fatigue life of alloy with high nitrogen content [25, 35]. I. Alvarez-Armas et al [36, 37] focused on assessing the damage evolution during low cycle fatigue (LCF) behavior in an embrittled duplex stainless steel type DSS S32205. After a grain-coarsening heat treatment the microstructure consists of approximately 50% austenite with a mean grain size of 30 μm embedded in 50% ferrite with a mean grain size of 27 μm. Finally, the material was aged at 475°C during 100h, resulting in Vickers hardness values of 254 HV in the austenite and 465 HV in the ferrite. The tests were carried out under plastic strain control with a fully reversed triangular wave at $\Delta\varepsilon_p = 3\times10^{-3}$. After LCF tests the surface damage is seen in Figure10. The deformation initiates in the austenite. As cycling proceeds, the slip markings intensify homogeneously in austenite while deformation is localized heterogeneously in ferrite. Finally, short cracks nucleate in the ferrite phase, Figure 10.

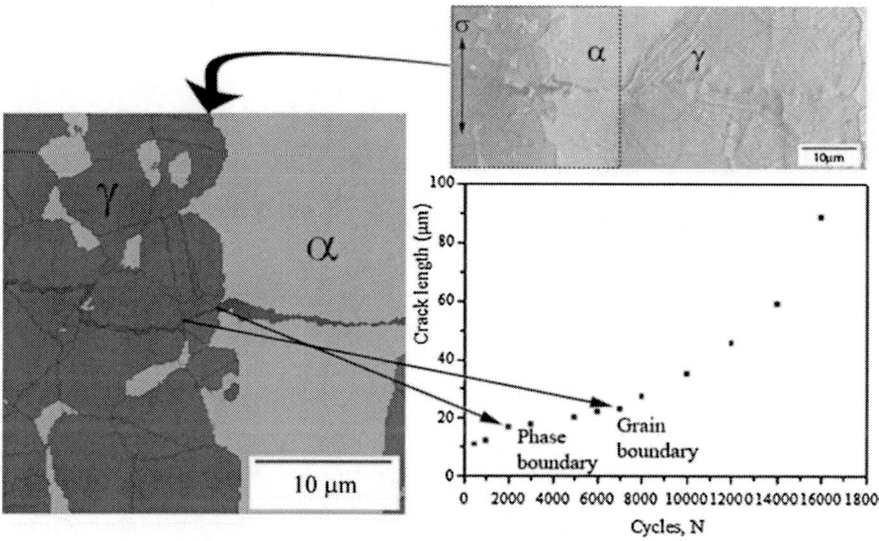

Figure 9. Superposition of SEM image and EBSD phase map showing the crack path after LCF test and the crack length vs. number of cycles [33].

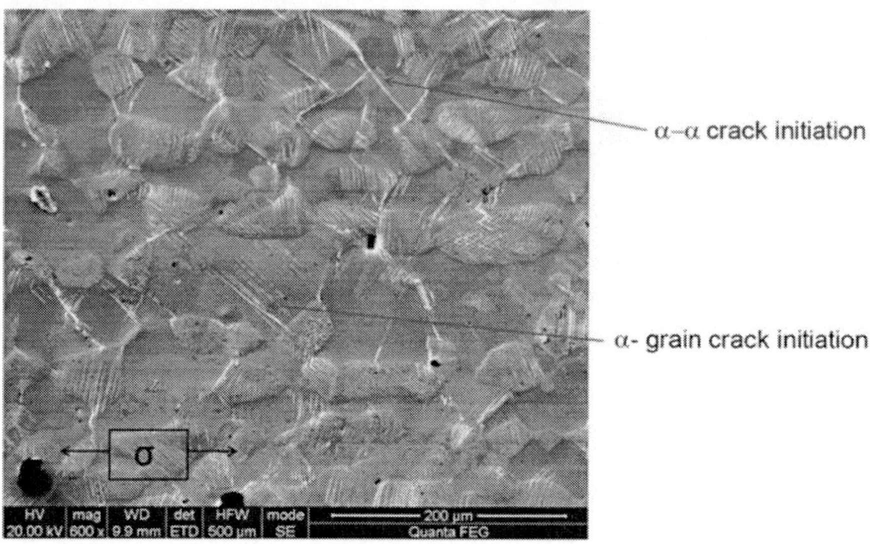

Figure 10. SEM micrograph showing the surface damage after low cycle fatigue [36].

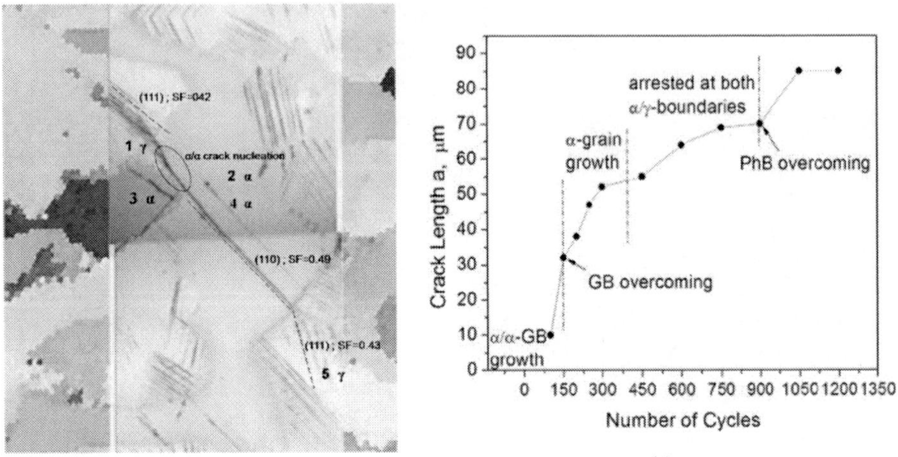

a) b)

Figure 11.Superimposed images of an EBSD grain map and the damage distribution micrograph recorded by the CCD camera. The inset diagram shows the micro-crack growth behavior [36].

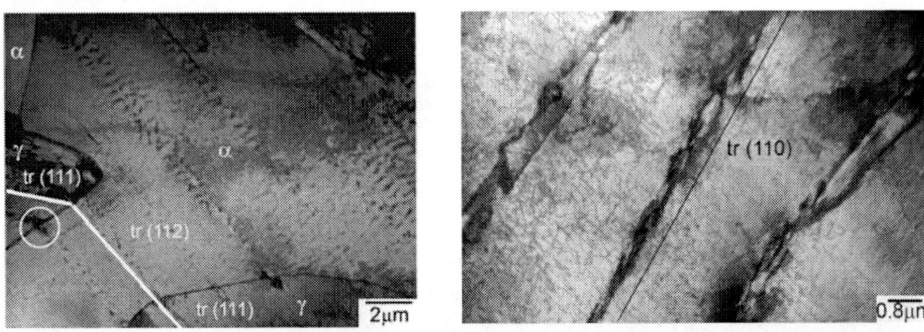

a) b)

Figure 12. Dislocation structure of ferrite at rupture: a) Dislocation bands. The specimen axis is in the vertical direction. The circle area indicates the deviation of the band as it crosses the γ/α boundary; [36]. b) microbands, observed near-surface region, developed in aged DSS S32750 [38].

The microcrack initiation and propagation is analyzed by the overlapping of a grain structure map obtained by EBDS and an image of the fatigue damage recorded by the CCD camera at rupture, Figure 11. In this figure, the traces of the activated slip planes with its corresponding Schmid factor (SF) are indicated. In all the cases, the planes with the highest SF are activated in both phases. Figure 11 shows a microcrack that nucleates in the α/α grain boundary (circled area, grains 2–3). The microcrack remains arrested at the α/γ phase boundary (grains 1–2), upper part of this figure, whereas it propagates along the lower α grain (grain 4) up to the second α /γ interphase grains (4–5). Finally, it remains arrested during several cycles and finally, both microcrack tips propagate into the upper (grain 1) and lower (grain 5) austenite grains, respectively. The inset diagram shows the microcrack growth behavior versus the number of cycles. It is evident that the propagation rate in the ferritic grain is very fast. The embrittled phase needs low energy to let a nucleated crack propagate along a crystallographic plane. The microcracks propagate fast in the ferritic grains, decelerate or remain stable at the phase boundary and accelerate again, but with a lower rate, having entered the austenitic grain.

During fatigue, these researchers observed that the austenitic grains deform in a planar way, with mainly two slip systems plastically active. Due to constraining effects caused by α' zones on dislocation in the aged ferrite, the plastic deformation is not easy. Hence, the dislocation structure has a less wavy slip character as usually occurs in a bcc structure and dislocations are more confined to their glide planes. Therefore, the plasticity is conducted along channels, either as dislocation bands or microbands, Figure 12. The crystallographic orientation of the dislocation band correspond to the activation of the most favorable slip planes either {110} and/or {112} according to the Schmid Factor (SF), Figure 12 a). Balbi et al reported [38], in aged DSS S32750 cyclic tested $\Delta\varepsilon_p = 3\times10^{-3}$, that the dislocations bands found in the bulk gives rise to the formation of a microband structure in near-surface of the specimen, Figure 12 b). Microbands (MB) are formed by a double dislocation wall separated approximately 0.2–0.4 μm and lays on the {110} planes and enclosing a misoriented region with respect to the surrounding matrix. Sometimes, microbands resemble elongated cell structures. MB's width remains nearly unchanged but theirs length, interior dislocation microstructure and number of MB within the same grain increase with increase in the strain amplitude and depend strongly on intrinsic grain orientation and deformation of neighboring grains. Hereñú et al [39] examined by energy dispersive x-ray spectroscopy (EDS) these microbands and reported that inside these microbands the composition fluctuation (Fe, Cr) due to the spinodal decomposition disappear during the fatigue life. Therefore, a soft α-phase is immersed in a matrix of α' hard –phase producing stresses on both sides of the interphase. Consequently, a dislocation arrangement at the α'/α interphase plane is formed during cycling. When the capacity of this α'/α interphase to plastically deform saturates, microcracks nucleate on them. If the microcrack arrives at the phase boundary, after propagating in the ferritic grain, it needs that the neighboring austenite grain hardness due to strain localization to continue growing. As plastic strain accumulation in austenitic grain could be great, the microcracks are arrested in this interphase.

3.2. HCF

3.2.1. DSS As-Received

A recent detailed research was conducted to study the surface damage and short crack behavior of DSS S32205, with enlarged grains by heat treatments, during HCF under stress control of $\Delta\sigma/2 = 350$ MPa [40, 41, 45]. Figure 13a) shows that both phases present cyclic plastic activity. Nevertheless, it is interesting to note that the concentration of stresses in the ferrite increases with the number of cycles and slip markings intensify and turn into coarse bands that could remain arrested at the grain boundary. The short cracks initiate mostly along these coarse slip bands in the ferrite. TEM observations give evidence of the plastic deformation sustained by both phases, Figure 13b). Planar arrangements and dislocation pile-ups in two slip systems are the characteristic dislocation structures observed in the austenite phase while the dislocation structure in the ferrite consists of loop patches of dislocations. It can be seen in Figure 13b) a dislocation band in the austenite that propagates in the neighbour ferritic grain where is not visible high stress concentration at the grain boundary and the dislocations can cross it easily. Zielinski et al. [42] studied the evolution of dislocation structure in an annealed DSS by in situ TEM straining experiments and reported that the evolution of dislocation structure during straining was dependent on the orientation relationship between the two phases. On the other hand, Taisne et al. [43] analyzed the role of interfaces in fatigue deformation mechanism in DSS bicrystal and stated that phase boundary geometry and elasticity affect the dislocation transmission process. According to Zhai et al. [44], the twist (α) and tilt (β) angles of the crack-plane deflection at a grain boundary are the key factors that control the path and growth rate of a short crack. The resistance to microcrack propagation will be smaller when the available slip planes in the next grain forms small α and β. Marinelli et al. [45] observed that when the α and β angles between slip markings in two neighboring austenite and ferrite are less than 10º, the plastic strains of the neighboring grains are compatible, and no stress concentration is developed at the phase boundary. On the contrary, if α is higher than 10º, the dislocations cannot cross the phase boundary and plastic deformation is accumulated in the first phase, usually the austenite, until a stress concentration occurs at the phase boundary.

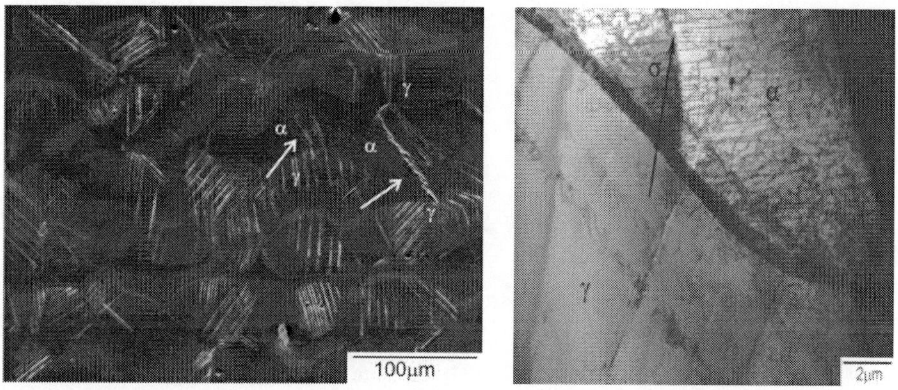

a) b)

Figure 13. a) Cyclic plastic activity in as-received DSS during HCF; b) dislocation pile-ups crossing a phase boundary [45].

Recently, Scharnweber et al [46] reported that the most frequently crack initiation sites of annealed DSS S32205 are ferritic grains boundaries followed by slip bands in austenite, phase boundaries and twin boundaries in austenitic grains (Figure 14). Hereñu et al [47] could rationalize the cracks initiation at ferritic grains boundaries. They found that thermal treatments at high temperature followed by water cooling produce Cr_2N precipitates in the ferritic phase of DSSs UNS S32205 and UNS S32750 that detrimentally affect their fatigue life. The thermal treatments carried out to enlarge the grain size [36, 40] may enhance the precipitation of needled Cr_2N leading to stress concentration that could cause premature crack initiation.

Strubbia et al [48] analized the HCF damage in LDSS AL 2003. TEM observations show that whereas dislocation structure of the ferrite phase is similar to that found before the fatigue testing (Figure 15a), the austenite phase presents pile-ups on one, two or three active slip systems (Figure 15b). Moreover, when the pile-ups in the austenite are arrested at a grain boundary produce a high dislocation density in this zone (Figure 16). It is relevant to remark that contrary to LCF, where short cracks nucleate along extrusions in ferrite, in this regime short cracks they nucleate at grains boundaries (Figure 17). The cracks at α/α boundaries could be attributed to asymmetric slip of screw dislocations which led to shape changes in neighbouring ferritic grains.

On the other hand, stress localization at grain boundaries could cause crack nucleation α/γ boundaries.

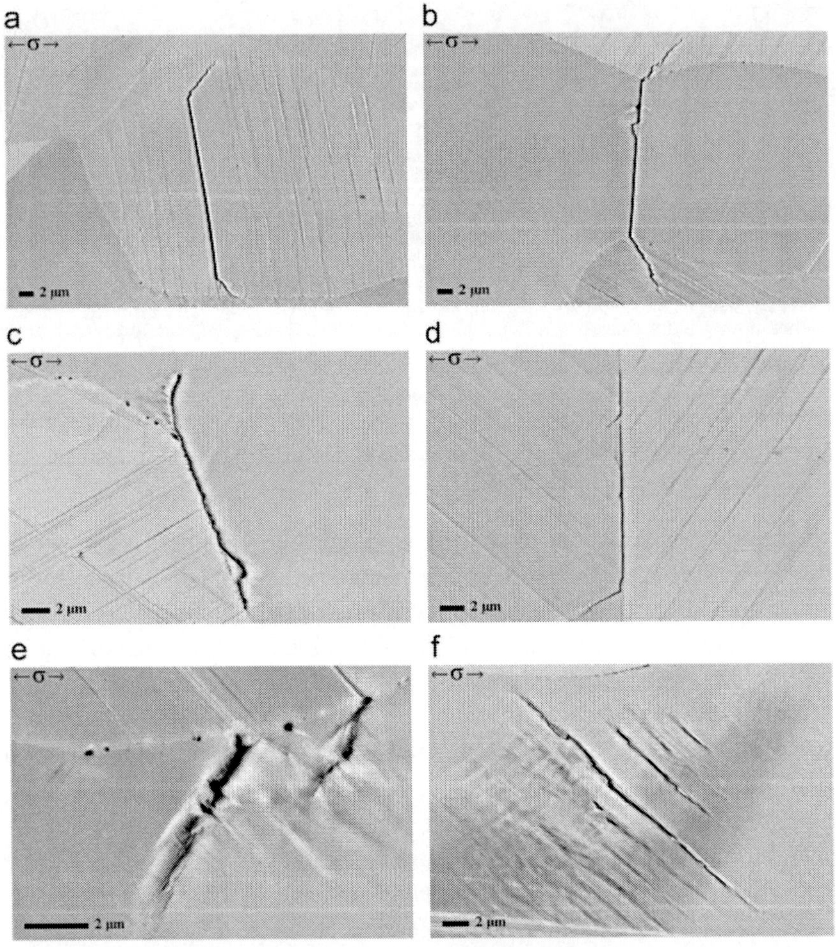

Figure 14. Crack initiation sites in the duplex steel investigated: (a) slip band in austenite, (b) grain boundary ferrite–ferrite, (c) phase boundary,(d) twin boundary in austenite, (e) slip bands in ferrite at α phase boundary and (f) slip bands in ferrite [47].

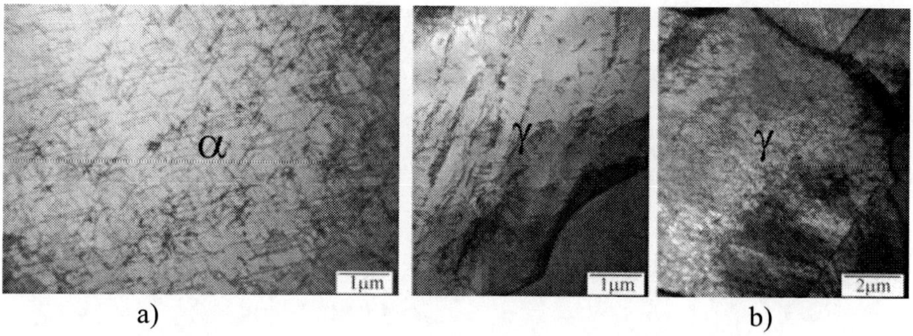

Figure 15. Dislocation microstructure in HCF specimens fatigued up to failure (a) ferrite (b) austenite [48].

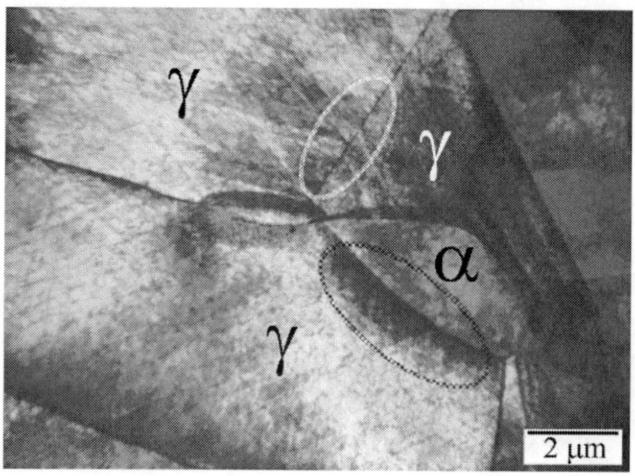

Figure 16. Zones of high stresses localized in grains boundaries observed in HCF specimens fatigued up to failure [48].

Examining the crack propagation in the DSS S2205, Figure 18a) shows a crack that is initiated in the ferrite on the $(1\bar{1}\bar{2})[1\bar{1}1]$ system with the highest SF= 0.47 and after a few cycles the right crack tip propagates into the austenite grain on the $(1\bar{1}\bar{1})$ $[110]$ slip system with SF= 0.21; this crack remains arrested in the next phase boundary during few cycles and finally surpass it and grows on the $(1\bar{1}0)[11\bar{1}]$ slip system in ferrite with SF= 0.44. On the other hand, the left crack tip grows up to the α/γ boundary where it remains arrested. This fact is shown as a plateau in Figure 18b). Therefore, it is possible to think that phase boundaries are an effective barrier against crack propagation because they delay the advance of the crack tip.

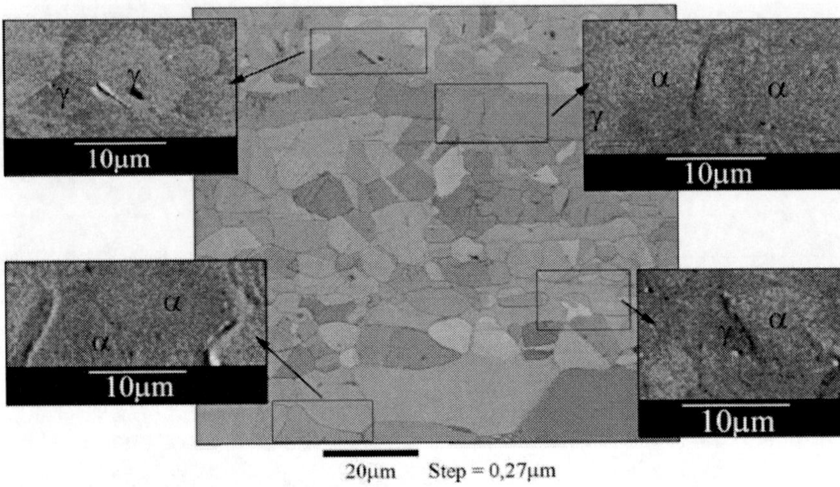

Figure 17. SEM image and EBSD map showing cracks nucleation in grain boundaries in HCF specimens [48].

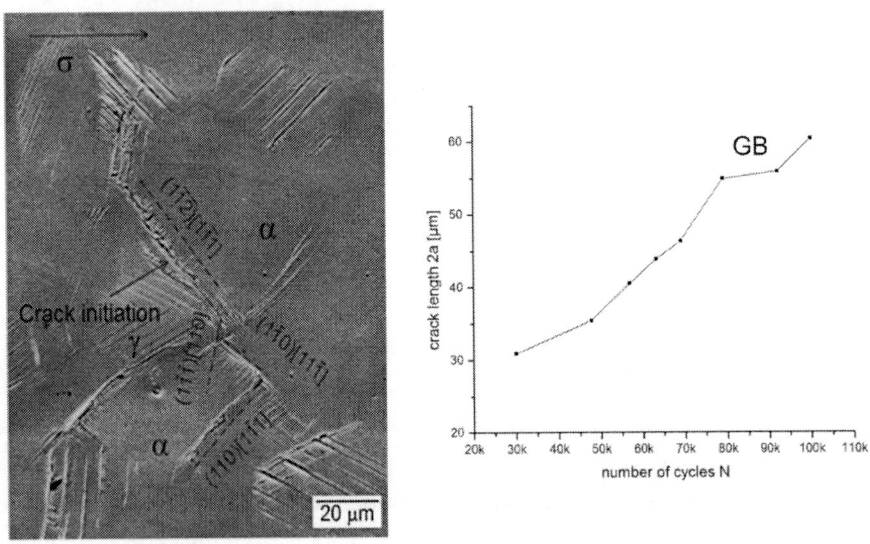

a) b)

Figure 18. Homogenized DSS S32205: (a) scanning electron micrograph showing the crack path after HCF at $\Delta\sigma/2$ 350 MPa and (b) crack length against the number of cycles [45].

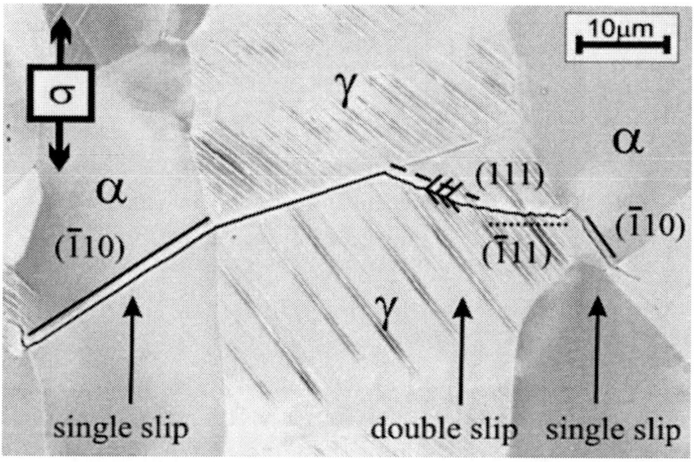

Figure 19. Single and double slip crack propagation [40].

According to Düber et al [40], short crack propagation in HCF regime can be subdivided into two basic mechanisms: operating either by single-slip (crystallographic propagation) and double-slip (propagation perpendicular to loading axis), giving rise to substantially different propagation rates. Additionally, it was found that a crack, which has grown in double slip, can change to the single slip mechanism again when entering a new grain and no adequate second slip plane is available, (Figure 19).

3.2.2. DSS Aged

Marinelli et al [41] and Alvarez et al [36] studied the effect of the 475ºC embrittlement on the fatigue limit and cracks propagation during HCF in a standard duplex stainless steel DSS S32205. The results showed that the fatigue limit of the embrittled samples is substantially larger than that of the conventional samples in homogenized condition (Figure 20). Moreover, they observed that the first slip markings appear mostly in the austenitic phase and as cycling proceeds, these intensify and some propagate into the neighboring ferritic grains or remain arrested at boundaries. According to a study [36, 49] microcrack behavior mainly follows two different situations: i) nucleation at the α/α boundary and propagation along the grain boundary up to the next α/γ boundary; ii) nucleation at the α/γ boundary and propagation in the α-grain up to the next barrier. Figure 21a) shows a typical microcrack behavior [45]. It initiates at the α/γ boundary and propagates on the left-hand crack tip, along the $(1\bar{1}\,\bar{2})[1\bar{1}1]$ system with SF= 0.46 in a ferritic grain, while on the right-hand crack tip the crack grows in the austenitic grain (black dashed line) by double slip on $(1\bar{1}\bar{1})[01\bar{1}]$ and $(1\bar{1}1)[110]$ systems with the highest Schmid factor, SF= 0.45 and SF= 0.41, respectively. As it is shown in Figure 21a), the direction of subsequent crack propagation can be represented by the vector addition of the displacements along the two slip systems. It is relevant to point out that phase boundaries represent an efficient barrier to crack propagation. This fact is represented by a plateau in the curve of Figure 21b). It is evident from this figure that the barrier effect of the phase boundaries is generally higher than the barrier effect of the ordinary α/α and γ/γ grain boundaries, irrespective of the angles between possible slip systems in the adjacent grains. In this sense, Llanes et al [50] studied the influence of thermal aging on the HCF resistance of UNS S31803 in terms of short crack behavior. They rationalized the higher fatigue limit and fatigue strength observed in aged DSS in comparison to the annealed DSS by the arrest of microcracks, nucleated in austenite, at interfases α/γ.

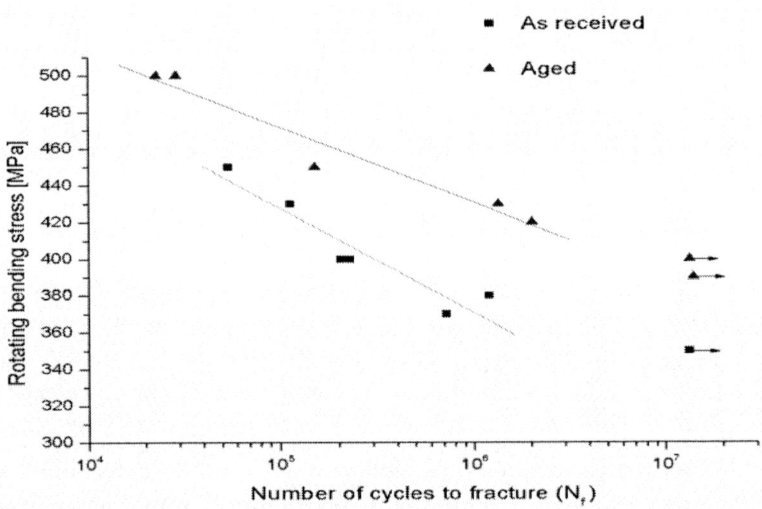

Figure 20. S/N (Wöhler) diagram of DSS S32205 for rotating bending tests in the homogenized and aged conditions [45].

Marinelli et al [41] also correlated the dislocation structure developed in the near surface under HCF with the aforementioned short crack behavior. These authors found that some grains of the ferrite are practically inactive and others show low dislocation density such as cross-stitch dislocation structure

(Figure 22a). However, where the pile-ups in the austenite impinge the phase boundary, the localization of cyclic strain causes the increase of the stress concentration near the phase boundary of the neighboring ferritic grain, creating a more developed dislocation structure as shown in Figure 22b). As a matter of fact, the high yield stress required to deform plastically the embrittled ferrite makes the austenite be the only phase capable to accommodate the deformation. Nevertheless, once the capacity of the austenite grain to plastically deform is reached, the pile-ups generate strong stress concentration zones at the phase boundary capable to allow the nucleation of microcracks and the activation of slip systems in the adjoining ferrite. Concerning crack initiation at α/α grain boundary, where the plastic activity is scarce, there is a strong effect of elastic anisotropy causing stress that can contribute to crack initiation [51]. This strong elastic stress-concentration may also produce micro cleavage of slip bands [52]. When microcracks propagation occurs inside the grains some difference between both phases should be accounted. Whereas microcracks propagate along a single-slip plane with the highest SF in the ferrite grains, in the austenite grains can also initially propagate across alternating slip planes and finally continue along a single slip plane with the highest SF.

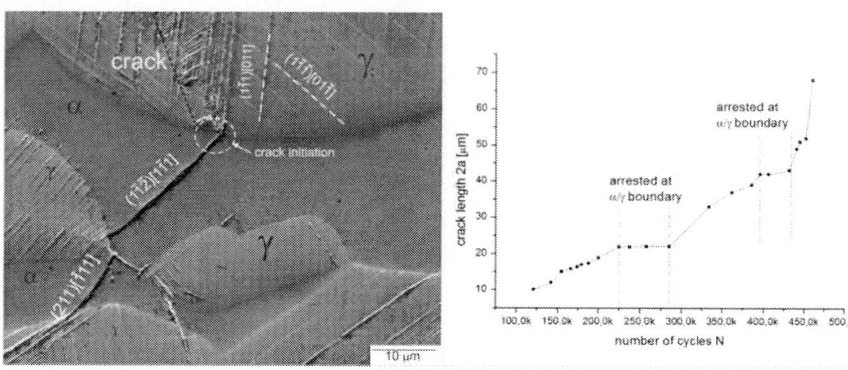

a)

b)

Figure 21. a) Scanning electron micrograph showing the crack path in HCF at $\Delta\sigma/2$= 350 MPa; b) crack length vs. the number of cycles [45].

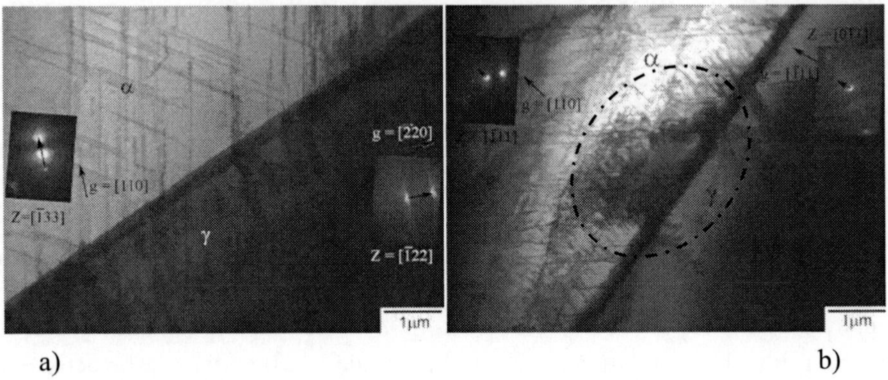

Figure 22. a) Cross stitch dislocation structure in ferrite; b) high stress concentration zones due to pile-ups of dislocations at grain boundary [45].

4. Crack Propagation Model

A short crack propagation model is used to identify the behavior of the cracks in the DSS during HCF and LCF. This model developed initially by Schick [53] and extended by Künkler et al. [54] is based in the model of Navarro and de los Rios published in 1988 [55,56]. In this model, plastic slip ahead of a growing microcrack is blocked by the grain boundary until the stress on a dislocation source in the neighboring grain exceeds a critical value. Then, the plastic deformation is transferred on the slip system in the adjacent grain and the crack propagation rate increases.

In order to model the microcracks behavior it is necessary to know the grain and phase distribution with the respective crystallographic orientations, the critical shear stress to activate plastic slip, and the barrier strength of phase and grain boundaries. EBSD measurements of surface area allow the reproduction of the respective microstructure. By knowing the direction dependent values of the stiffness tensor (elastic anisotropy) and the directions of the slip systems of the grains involved, locations of plastic strain concentration were predicted. These locations were correlated with the slip band and/or crack initiation sites. Once a crack nucleates, it is assumed that it propagates along slip bands This process has been assessed by a boundary-element (BEM) approach [54, 56, 57], where the crack and the adjacent slip bands are meshed by boundary elements that represent tangential displacement steps within the slip band and tangential/normal displacement steps within the crack area. These displacement steps can be considered as mathematical dislocation dipoles. The crack elements (i) consist of two dislocation dipoles with a Burgers vector parallel and normal to the crack to allow for sliding and an opening displacement of the crack flanks, respectively. The slip bands ahead of the crack are restricted to tangential displacements (assuming elastic ideal-plastic material behavior), so the respective elements (j) only consist of a dislocation dipole parallel to the crack. The elements are connected to each other by influence functions G_{ij}, which describe the stress in an element i due to a displacement in an element j, [56]. The total stress σ^{i}_{nn} and τ^{i}_{tn} are obtained by a summation over all elements (p crack elements and q elements within the slip bands ahead of the crack tip) plus the external normal and shear stresses $\sigma^{i\infty}_{nn}$ and $\tau^{i\infty}_{tn}$. For all slip band and crack elements an equation system (Eq. 1) can be setup and solved accounting for the boundary conditions that(i) neither tensile normal stresses or shear stresses may occur within the crack area, (ii) the normal displacement b_n in the crack area must always be positive, and (iii) within the plastic zones ahead of the crack tip the shear stress τ must not exceed the resistance τ_b to dislocation motion (assumed as to be equal to the microstructural cyclic yield stresses [40]).

$$b_{\mathrm{n}} \geq 0 \qquad i = 1 \ldots p$$

$$\sigma^{i}_{\mathrm{nn}} = \sum_{j=1}^{p} G^{ij}_{\mathrm{nn,n}} b^{j}_{\mathrm{n}} + \sum_{j=1}^{p+q} G^{ij}_{\mathrm{nn,t}} b^{j}_{\mathrm{t}} + \sigma^{i}_{\mathrm{nn,a}} \leq 0 \qquad i = 1 \ldots p$$

$$\left|\tau^{i}_{\mathrm{tn}}\right| = \left|\sum_{j=1}^{p} G^{ij}_{\mathrm{tn,n}} b^{j}_{\mathrm{n}} + \sum_{j=1}^{p+q} G^{ij}_{\mathrm{tn,t}} b^{j}_{\mathrm{t}} + \tau^{i}_{\mathrm{tn}}\right| \begin{cases} = 0 & i = 1 \ldots p \\ \leq \tau_b & i = p+1 \ldots p+q \end{cases} \tag{1}$$

Solution of the equation system provides the displacement field along the crack and the yield strips (slip bands) in front. The displacement at the crack tip, i.e., displacement interval between maximum and minimum load ΔCTSD is used as crack driving force, using the following crack propagation law:

$$\frac{da}{dN} = C \cdot \Delta CTSD \tag{2}$$

with C being a constant representing the irreversible fraction of crack-tip displacement. This constant was adapted to the experimental results yielding C

= 0.002, which was used for all simulations in the embrittled steel during HCF [45], and C = 0.08 for LCF [58]. Slip transmission from the first grain to the neighbouring grains is considered by the use of sensor elements, which are placed at the grain boundaries, representing the potential slip systems of the respective grains (Figure 23). During each calculation step, crack advance results in a decrease in ΔCTSD along the slip band in grain 1 and a stress increase at the sensor elements, respectively. As soon as a critical stress value is exceeded at one of the sensor elements, the respective slip band is meshed by a set of boundary elements and the displacement field is extended by one grain.

This model takes the crack propagation operating by single and double-slip mechanism into account, i.e., the crack can grow on one slip system resulting in straight crack paths which follow a slip plane with a high Schmid factor or the crack path is the result of two alternating operated slip systems [59]. This model was successfully verified for fatigue crack propagation in homogenized [40] and embrittled DSS [45, 58]. Figure 24 and Figure 25 show the crack path and the simulated crack path for embrittled DSS during LCF and HCF respectively. In both situations the experimentally observed cracks were defined in the model using the geometry and crystallographic orientation of the respective grains. The length of the additionally activated slip planes is restricted by the grain boundaries. Moreover, starting cracks were defined according to the experimentally observed ones. Then, these cracks were subjected to cyclic loading calculations.

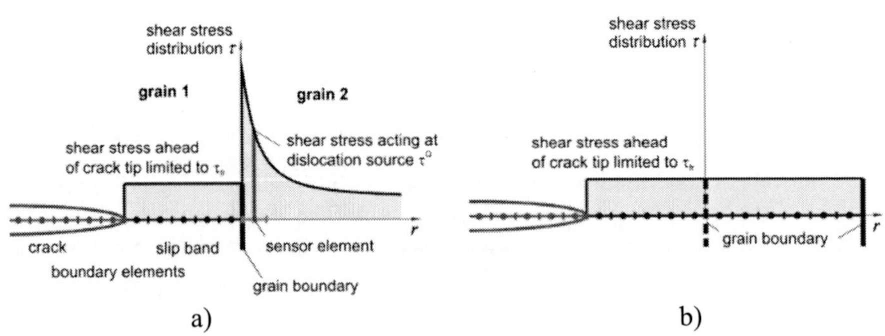

Figure 23. Schematic representation of the boundary element modelling concept showing the stress distribution within the yield strip ahead of the crack tip: (a) plastic slip blocked by grain boundary, and (b) slip transmission to neighbouring grain [58].

Due to the fact that the LCF tests were carried out under plastic strain control, the stress amplitude used for calculation was that attained during the saturation stage and corresponds too $\Delta\sigma/2 = 300$ MPa.

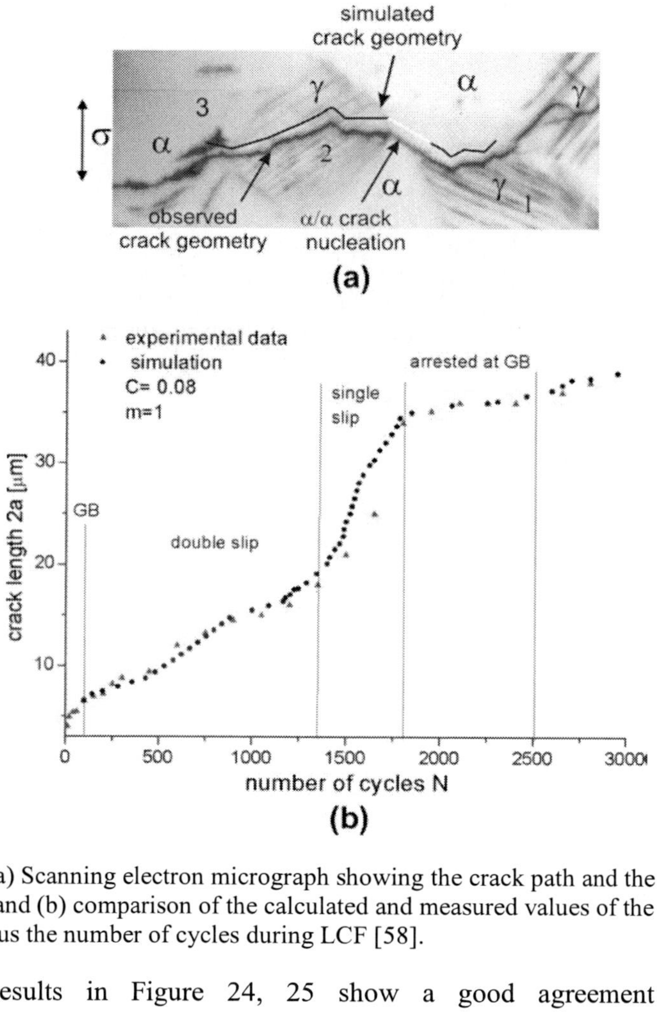

Figure 24. (a) Scanning electron micrograph showing the crack path and the simulated crack path; and (b) comparison of the calculated and measured values of the crack lengths versus the number of cycles during LCF [58].

The results in Figure 24, 25 show a good agreement between experimentally observed and simulated crack propagation. Moreover, the blocking of the slip bands at the grain boundaries as well as the retardation of the crack growth in the vicinity of the α/γ phase boundaries, caused by pile-up of dislocations, could be qualitatively simulated. Some authors [60, 61] have reported that the crack propagation rate is strongly influenced by the misorientation of the crack and the slip-direction between grains. Lillbacka et al [60] have studied the cracks behavior in a duplex steel and they reported that if the slip-directions of a grain were rotated such that the evolution of plastic slip became smaller than for the neighboring grain, the crack propagation rate decreased as the crack approached the grain boundary.

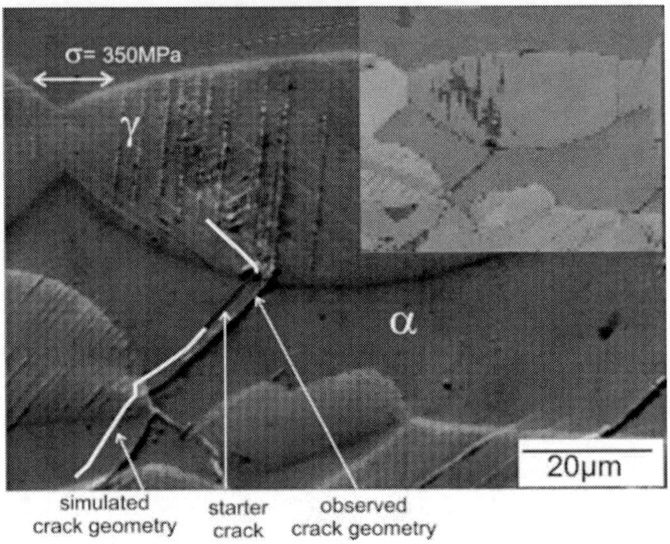

a)

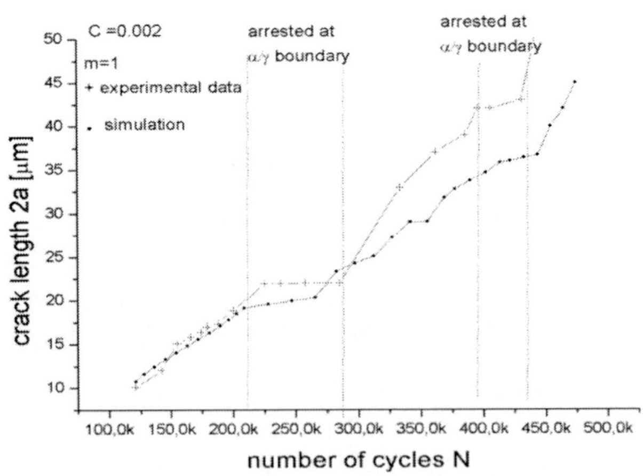

b)

Figure 25. Aged 1.4462 DSS: a) scanning electron micrograph showing the crack path; b) comparison of the calculated and measured values of the crack lengths against the number of cycles during HCF.

Evaluation of crack paths on the surface of DSS revealed that cracks in the austenite phase tend to grow along two alternating slip bands while in the ferrite phase the crack mostly grows on a single-slip plane. Of course, microcrack propagation cannot be attributed only to the effect of ΔCTSD on two-dimensional slip planes. Additionally, three-dimensional crack propagation into the bulk must be taken into account. However, the calculated crack propagation rates for different cracks are in good agreement with the experimental results. Düber et al. [40] argued that the reason for this good agreement is the fact that short crack propagation is governed mainly by the microstructure of the surface grains due to plane stress conditions.

CONCLUSION

The present research analyzes the damage produced by cyclic loading in DSS from its early stages by observation of the evolution of surface relief to the final short crack initiation and growth. The relation of this fatigue damage with the dislocation structure is also attempted.

In DSS the nature of short crack nucleation sites and consequently that of structural barriers depends on numerous microstructural parameters and also on the respective response of each phase with respect to testing conditions, ie LCF or HCF. The effectiveness of a barrier depends of orientation and the degree of deformation developed in the neighbouring grain. In the aged DSS, due to the strong hardening of the ferritic phase, plastic activity is exclusively concentrated within the more ductile austenitic grains. Therefore, in this case the fatigue damage also depends on maximum plastic deformation sustains by the austenitic phase which determines the resistance of the interphase barriers.

ACKNOWLEDGMENTS

The financial support by Deutsche Forschungsgemeinschaft DFG, Deutscher Akademischer Austauschdienst (DAAD), Alexander von Humboldt Fundation and Consejo Nacional de Investigaciones Científicas y Técnicas (CONICET Argentina) are gratefully acknowledged.

REFERENCES

[1] K. H. Lo, C.H. Shek, J. K. L Lai, Recent developments in stainless steel; *Materials Science and Engineering* R65 (2009) 39-104.

[2] I. Alvarez-Armas, Duplex Stainless Steels: Brief History and Some Recent Alloys, *Recent Patents on Mechanical Engineering* (2008) 1, 51-57.

[3] N. Giel, Application of duplex stainless steel in the chemical process industry, *Proc. 5th World Duplex Stainless Steel Conference* 1997, Maastricht, The Netherlands, KCI Publishing D97-201 (1997) 9-16.

[4] L. Smith, A guideline to the successful use of duplex stainless steels for flowlines. *Proc.Duplex America Conference*, Houston, USA, KCI Publishing DA_2-102 (2000) 17-29.

[5] N. Johan, Digesters and pulp storage towers of duplex stainless steels-saving weight and costs, ACOM, *Avesta Sheffield* AB 3 (1994) 1-8.

[6] M. Snisand, J. Olsson, Reduce costs for storage and distribution of desalted water -- use duplex stainless steel. *Desalination* (2008) 223(1-3) 476-486.

[7] J. Da-Wei, G. Chang-Sheng, Z. Xiang-Juan, L. Junl, S. Lu-Lu, X. Xue-Shan, 22Cr high-Mn-N low-Ni economical duplex stainless steels, *Journal of Iron and Steel Research, International.* (2012) 19(2): 50-56.

[8] E Alfonsson. Lean duplex – the first decade of service experience. In: *8th Duplex Stainless Steels Conference*. Beaune, France: 2010.

[9] R.O. Williams, H.W. Paxton, The nature of ageing of binary iron-chromium alloys around 500°C, *J. Iron Steel Inst.* 185 (1957) 358-374.

[10] L. Llanes, A. Mateo, L. Iturgoyen, M. Anglada, Aging effects on the cyclic deformation mechanism of a duplex stainless steel, *Acta Mater.* 44 (1996), pp. 3967–3978

[11] K. L. Weng, H. R. Chen, J. R. Yang, The low-temperature aging embrittlement in a 2205 duplex stainless steel, *Materials Science and Engineering* A 379 (2004) 119–132.

[12] T. Otarola , S. Hollner, B. Bonnefois, M. Anglada, L. Coudreuse, A. Mateo, Embrittlement of a superduplex stainless steel in the range of 550–700°C, *Engineering Failure Analysis* 12 (2005) 930–941.

[13] A. F. Armas, S. Degallaix, G. Degallaix, S. Hereñú, C. Marinelli, I. Alvarez-Armas, Beneficial effects induced by high temperature cycling in aged duplex stainless steel, *Key Engineering Materials Vols.* 345-346 (2007) pp. 339-342.

[14] H. Mughrabi, R. Wang, K. Differt, U. Essmann, Fatigue crack initiation by cyclic slip irreversibilities in high cycle fatigue. In Fatigue Mechanisms; *ASTM STP* 811 (1979) 5-45.

[16] P. Lukas, L. Kunz, Cyclic plasticity and substructure of metals, *Materials Science and Engineering* A 322(2002) 217-227

[17] J. Polak, Mechanisms and kinetics of the early fatigue damage in crystalline materials, *Materials Science and Engineering* A 468-470 (2007) 33-39.

[18] P. Lukas, L. Kunz, Cyclic plasticity and substructure of metals, *Materials Science and Engineering* A 322 (2002) 217-227.

[19] P. Broz, Metal fatigue features, Acta Polytechnica 46 (2006) 34-41.

[20] J. Man, K.Obrtlik, J. Polak, Extrusions and intrusions in fatigued metals. Part 1. State of the art and history.*Phil Mag* (2009) 89:1295-336.

[21] J. Polak, T. Kruml, K. Obrtlik, J. Man, M. Petrenec. Short crack growth in polycrystalline materials, *Procedia Engineering* 2 (2010) 883–892.

[22] S. Suresh, *Fatigue of materials,* 2nd Edition (1998), Cambridge University Press

[23] A. El Bartali, V. Aubin and S. Degallaix, Fatigue damage analysis in a duplex stainless steel by digital image correlation technique Fatigue Fract. *Engng. Mater Struct.* 31, (2007) 137–151.

[24] J. B. Vogt, D. Salazar, I. *Proriol Serre. Duplex Stainless Steel,* Chapter 8, Editorial Wiley (2009).

[25] M. Balbi, M. Avalos, A. El Bartali, I. Alvarez-Armas, Microcrack growth and fatigue behavior of a duplex stainless steel, *International Journal of Fatigue* 31 (2009) 2006–2013

[26] W. Zielinski, W.Swiatnicki, M. Barstch, U. Messerschmidt, Non-uniform distribution of plastic strain in duplex steel during TEM in situ deformation. *Mater. Chem. Phy.* (2003) 81, 476–479.

[27] D.A. Lott-Crumpler, *Modelling Simul. Mater. Sci. Eng.,* Vol 5 (1997) 317-36.

[28] M. C. Marinelli , S. Degallaix, and I. Alvarez-Armas, Short crack initiation during low-cycle fatigue in SAF 2507 Duplex Stainless Steel. *Key Engng Mater.* (2007) 345–346, 343–346.

[29] M.C. Marinelli, M.G. Moscato, J.W. Signorelli, A. El Bartali, I. Alvarez-Armas. K–S relationship identification technique by EBSD. *Key Engng. Mater.*(2011);465:415–8.

[30] M.C Marinelli, A. El Bartali , J. W. Signorelli , P. Evrard, V. Aubin, I. Alvarez-Armas, S. Degallaix-Moreuil, Activated slip systems and

microcrack path in LCF of a duplex stainless steel, *Materials Science and Engineering* A 509 (2009) 81–88.

[31] H. Mughrabi, *Proc 50th Anniv. Concept of Dislocations in Crystals,* p. 244. Inst. of Metals London (1985).

[32] I. Gutierrez and Amaia Iza-Mendia in: *Duplex Stainless Steels,* edited by I. Alvarez-Armas and S. Degallaix Moreuil, Wiley-ISTE, London, UK (2009).

[33] R. Strubbia, S. Hereñú, A. Giertler, I. Alvarez-Armas, U. Krupp, Experimental characterization of short crack nucleation and growth during cycling in lean duplex stainless steels, *Int. J Fatigue* (2013), http://dx.doi.org/10.1016/j.ijfatigue.2013.08.023

[34] R. Lillbacka, G. Chai, M. Ekh, P. Liu, Johnson E, Runesson K. *Acta Mater* (2007);55(16):5359–68.

[35] A. F Armas, S. Hereñú, I. Alvarez-Armas, S. Degallaix, A. Condó, F. Lovey, The influence of temperature on the cyclic behavior of aged and unaged super duplex stainless steels. *Mater. Sc. Eng A (*2008) 491:434-39.

[36] I. Alvarez-Armas, U. Krupp, M. Balbi, S. Hereñú, M.C. Marinelli, H. Knobbe, Growth of short cracks during low and high cycle fatigue in a duplex stainless steel, *International Journal of Fatigue* 41 (2012) 95–100.

[37] Alvarez-Armas I, Knobbe H, Marinelli MC, Balbi M, Hereñú S, Krupp U. Experimental characterization of short fatigue crack kinetics in an austenoferritic duplex steel. *Pro. Eng* (2011) 10: 1491–6.

[38] M. Balbi, S. Hereñú, I. Proriol Serre, J.-B. Vogt, A. F. Armas, I. Alvarez-Armas, Initiation and growth of short cracks during cycling in an aged superduplex stainless steel, *Key Eng. Mater* (2011) 465:411–4.

[39] S. Hereñú, M. Sennour, M. Balbi, I. Alvarez-Armas, A. Thorel, A.F. Armas, Influence of dislocation glide on the spinodal decomposition of fatigued duplex stainless steels, *Materials Science and Engineering A* 528 (2011) 7636– 7640.

[40] O. Düber, B. Künkler, U. Krupp, H. J. Christ, C.P. Fritzen. Experimental characterization and two-dimensional simulation of short-crack propagation in an austenitic–ferritic duplex steel*, Int J Fatigue* (2006) 28:983–92.

[41] M. C. Marinelli1, A.Giertler, J. Kumar Sahu, S. Hereñú, I. Alvarez-Armas, U. Krupp, *The Effect of Ferrite Embrittlement in Duplex Steel on Fatigue Crack Propagation from the Low (LCF) to the Very High Cycle Fatigue* (VHCF) Regime, ECF19, August 26-31. 2012, Kazan, Russia

[42] W. Zielinski, W. Witnicki, M. Barstch, U. Messerschmidt. Non-uniform distribution of plastic strain in duplex steel during TEM in situ deformation. *Mater Chem Phys* (2003);81(2–3):476–9.

[43] A. Taisne, B. Decamps, L. Priester. Role of interfaces in duplex stainless steel deformation micromechanisms. *Compos Interfaces* (2006);13(1):89–102.

[44] T. Zhai, A.J. Wilkinson, J.W. Martin. A crystallographic mechanism for fatigue crack propagation though grain boundaries. *Acta Mater* (2000);48:4917–27.

[45] M.C. Marinelli, U. Krupp, M. Kübbeler, S. Hereñú, I. Alvarez-Armas. The effect of the embrittlement on the fatigue limit and crack propagation in a duplex stainless steel during high cycle fatigue, *Eng. Fract Mech* (2013) 110: 421–429

[46] M. Scharnweber, W.Tirschler, C.-G.Oertel, W.Skrotzki. Initiation and propagation of short cracks in austenitic–ferritic duplex steel, *Materials Science&Engineering A* 583(2013)52–60

[47] S. Hereñú, M.G. Moscato, I. Alvarez-Armas, A.F. Armas. The influence of chromium nitrides precipitation on the fatigue behavior of duplex stainless steels. *Procedia Engineering 74 (2014) 179 – 182*

[48] R. Strubbia, S. Hereñú , A. Giertler , I. Alvarez-Armas , U. Krupp. Experimental characterization of short crack nucleation and growth during cycling in lean duplex stainless steels. *Int J Fatigue* (2013), http://dx.doi.org/10.1016/j.ijfatigue.2013.08.023

[49] U. Krupp, A. Giertler, M.C. Marinelli, H. Knobbe, H.J. Christ, P. Köster, Efficiency of grain and phase boundaries as microstructural barriers during HCF and VHCF loading of austenitic–ferritic duplex steel. In: Berger C, Christ HJ, editors. *Fifth international conference on very high cycle fatigue.* Berlin: DVM; 2011. p. 127–32.

[50] L. Llanes, A. Mateo, P. Violan, J. Méndez, M. Anglada, On the high cycle behavior of duplex stainless steel: Influence of thermal aging, *Materials Science and Engineering* A234-236 (1997) 850-852

[51] A. Heinz, P. Neumann. Crack initiation during high cycle fatigue of an austenitic steel. *Acta Metall Mater* (1990);38(10):1933–40.

[52] TJ Marrow, EJ King. Fatigue crack propagation mechanisms in a thermally aged duplex stainless steel. Mater Sci Eng A (1994);183:91

[53] A. Schick. *Ein neues Modell zur mechanismenorientierten Simulation der mikrostrukturbestimmten Kurzrissausbreitung,* vol. 18. Düsseldorf: VDI Verlag Reihe; (2004). p. 292.

[54] B. Künkler, O. Düber, P. Köster, U. Krupp, CP Fritzen, HJ Christ. Modeling of short crack propagation – transition from stage I to stage II. *Engng Fract Mech* (2008);75:715–25.

[55] A. Navarro, ER. de Los Rios. Short and long fatigue crack growth: a unified model. Philos Mag A (1988);57:15–36.

[56] U. Krupp. *Fatigue crack propagation in metals and alloys.* Weinheim: Wiley-VCH; 2007

[57] B. Künkler. *Mechanismenorientierte Lebensdauervorhersage unter Berücksichtigung der Mikrostruktur.* VDI-Fortschritt-Berichte, Reihe 18, Nr. 312, VDI Verlag, Düsseldorf, Germany; 2007.

[58] U Krupp, I Alvarez-Armas. Short fatigue crack propagation during low-cycle, high cycle and very-high-cycle fatigue of duplex steel – An unified approach. *Int J Fatigue* (2014) 65: 78–85

[59] B Künkler, CP Fritzen, O Düber, U Krupp, HJ Christ. Simulation of short crack propagation – transition from Stage I to Stage II. *PAMM Proc Appl Math Mech 2005*;5:341–2.

[60] R Lillbacka, E Johnson, M Ekh. A model for short crack propagation in polycrystalline materials. *Engng Fract Mech* (2006);73:223–32.

[61] G.P Potirniche, S.R Daniewicz. Finite element modeling of microstructurally small cracks using single crystal plasticity. *Int J Fatigue* (2003);25:877–84.

In: Recent Trends in Fatigue Design
Editor: Ricardo Branco
ISBN: 978-1-61668-410-5

Chapter 7

THE ROLE OF MICROSTRUCTURE ON FATIGUE CRACK PROPAGATION

Anghel Cernescu [*,1,2]
[1]Department of Applied Mechanics, Aalto University of Finland, Finland
[2]Mechanics and Strength of Materials Department, Politehnica University of Timisoara, Romania

ABSTRACT

The fatigue crack growth rate is the essential element of the evaluation concept based on damage tolerance and this made that the fatigue crack propagation to be the subject of many studies. The microstructure has fundamental influences on fatigue crack propagation. In this paper it is conducted a review based on experimental observations which aims to highlight the role of microstructure on the fatigue crack growth rate. It was observed that the presence of a hard constituent in ferrite matrix may have significant effects on the fatigue crack growth rate. Both morphology of hard constituent and volume fraction can determine the changing of propagation constants and introduces the retardation effects of fatigue crack. On the other hand the crystallographic characteristics and orientation of grain boundaries also influence the fatigue crack propagation. The twist and tilt angles of the slip systems from two adjacent grains may lead to retardation effects on crack

[*] Corresponding Author address. Email: anghelcernescu@gmail.com.

propagation, as well as the high-angle grain boundaries also may constitute barriers to fatigue crack propagation.

Keywords: Fatigue crack growth rate, pearlite, martensite, bainite, grain size, tilt and twist crack angles

INTRODUCTION

The fatigue life of a structure is generally based on two concepts: the knowledge of the structure itself and the knowledge of the loading mode of the structure. The procedures for estimating the lifetime had evolved over the time as a result of lessons learned from the failures recorded in engineering. Approaching the concept of safe operation by estimating the lifetime, was developed as response to the loss of some commercial and military aircrafts, although they were designed to meet static strength requirements. The concept of safe operations tries to maintain the safety of components by designing so that to have a higher fatigue strength than the estimated useful life. In this approach, the replacement time of a component estimated or determined by experimentally S-N curves, is based on the assumption that the item is free from defects. In the case of a component which has a defect resulted from the processing and maintenance conditions, the service life of that component may be less than the replacement time indicated by the fatigue tests. Equally a component may have a longer lifetime than the replacement period indicated by the fatigue tests. Because the lifetime of the components resulted from fatigue tests is divided by the large safety factors to obtain the replacement period, many components are replaced even though they do not exhibit any evidence of cracking by fatigue. As the approximations of safe operation concept become more evident, in the early 70s, [1], United States Air Force adopted a new method of defining the failure of components based on defects tolerance (damage tolerance concept). The concept based on defects tolerance assumes that each component has a potential defect and the fatigue life is based on the number of loading cycles required for a crack initiated to grow up to critical value. The fatigue crack propagation is controlled by the maximum normal stress and occurs by a repeated process of deformation and fracture in the vicinity of the crack tip. The propagation of fatigue cracks results from the formation, growth and coalescence of microcracks in a continuous process which is carried out in the main fatigue crack front. The microcracks can be formed at the main crack tip or ahead of it and they may not be coplanar with the main fatigue crack. Therefore, in accordance with experimental observations, the crack front itself is viewed as a sum of micro-fronts with positions and observations dependent on both the geometry of the body and the loading, as well as the size and orientation of grains and the crack propagation resistance of grain boundaries. Usually the increasing of crack length determines the acceleration of crack propagation.

The evolution of fatigue crack during its propagation can be followed based on the variation curve between crack length, a, and number of loading cycles, N, Figure 1. The basic parameter which appreciates the fatigue fracture is fatigue crack growth rate, da/dN, and represents the crack length increment during one loading cycle. Moreover, the fatigue crack growth rate represents the slope of the variation curve a = f(N) and can be written as:

$$\frac{da}{dN} = \frac{a_{n+1} - a_n}{N_{n+1} - N_n} \tag{1}$$

In general, the fatigue crack growth rate depends on the initial crack length and the amplitude or level of applied stress. However, these variables enter in the expression of stress intensity factor and based on this it was developed the idea that the fatigue crack growth rate can be correlated with stress intensity factor range, ΔK. The correlations between fatigue crack growth rate, da/dN, and stress intensity factor range, ΔK, are represented in the logarithmic scale as in Figure 2.

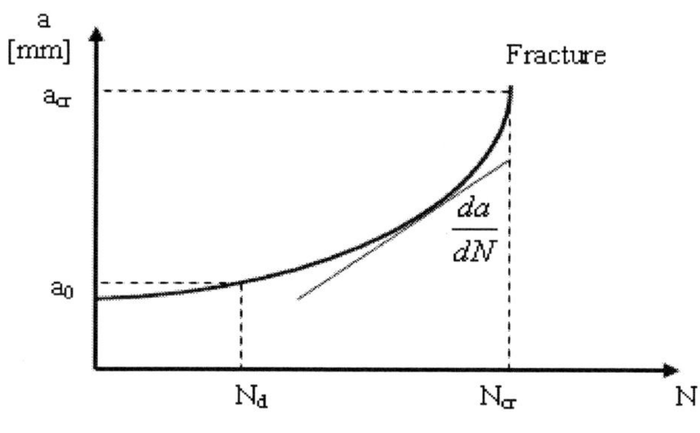

Figure 1. The variation curve between fatigue crack length and number of applied cycles: a_0 – the initial detectable crack length; a_{cr} – the critical crack length; N_d – the number of cycles corresponding to detectable crack length; N_{cr} – the number of cycles corresponding to critical crack length; da/dN – the fatigue crack growth rate.

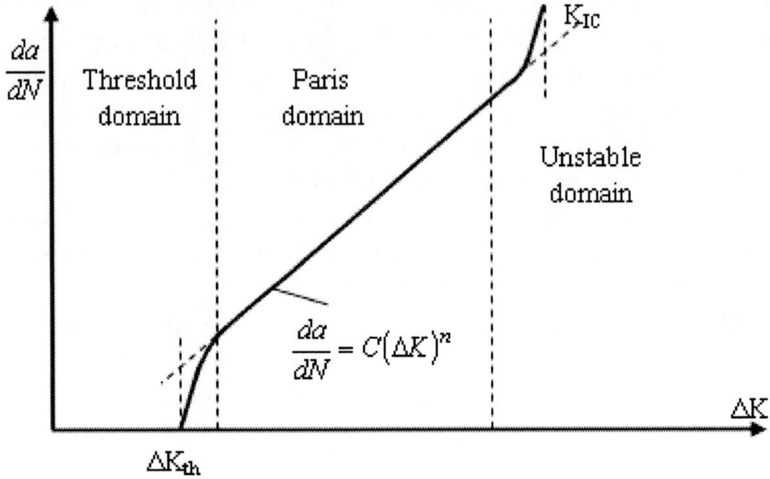

Figure 2. The variation curve of fatigue crack growth rate function of stress intensity factor range: da/dN – fatigue crack growth rate; ΔK – stress intensity factor range; ΔK_{th} – threshold stress intensity factor range; K_{IC} – fracture toughness.

To describe the variation curve da/dN = f(ΔK) different equations and analytical models have been developed. Among them the most known relationship is the so called "Paris' equation" which describes the variation of fatigue crack growth rate function of stress intensity factor range in the stable propagation domain (Paris domain) and has the following form:

$$\frac{da}{dN} = C \cdot \Delta K^n \tag{2}$$

where C and n are propagation constants.

Another well-known equation that can describe the variation curve da/dN = f(ΔK) in all three domains is NASGRO equation [2], that takes into account the effect of crack closure and has the following form:

$$\frac{da}{dN} = C\left(\frac{(1-f)}{(1-R)}\Delta K\right)^n \frac{\left(1-\frac{\Delta K_{th}}{\Delta K}\right)^p}{\left(1-\frac{K_{\max}}{K_c}\right)^q} \tag{3}$$

where da/dN is the crack growth rate, ΔK is the applied stress-intensity factor range, R is the stress ratio, ΔK_{th} is the fatigue threshold, K_{max} is the stress-intensity factor corresponding to peak applied load, K_C is the critical stress intensity factor, C and n are propagation constants, p and q control the shape of the asymptotes in the threshold and unstable domain and f is Newman's crack opening function [3].

In general all the fatigue crack growth equations contain the propagation constants C and n which are the defining parameters in predicting the relationship between da/dN and ΔK.

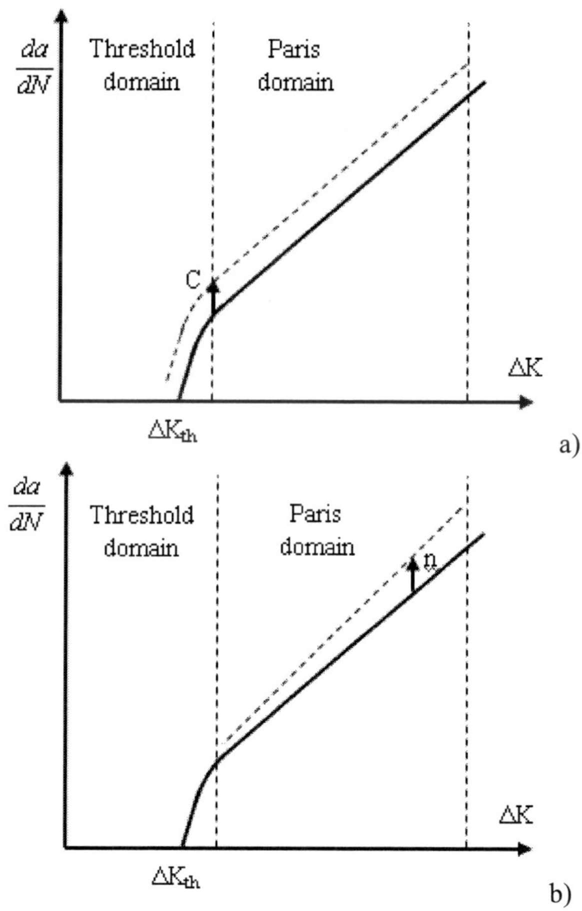

Figure 3. The propagation constants: a) the multiplicative constant C; b) the slope n.

It is known that the fatigue crack growth rate is directly influenced by the mean stress and stress ratio of the applied load, microstructure and working environment. The effect of these factors directly reflects on the propagation constants. For example, with increasing the mean stress and stress ratio, $R = \sigma_{min}/\sigma_{max}$, increases the fatigue crack growth rate by increasing the propagation constants C and n. Instead the microstructure has a large influence on fatigue crack propagation in the threshold domain and decreases as the fatigue crack growth rate increases and reaches the Paris domain.

The constant n is the slope of the fatigue crack propagation curve in the Paris domain and C is a multiplicative constant and has influence on the threshold domain, Figure 3.

Considering this it is possible to say that the microstructure has a greater influence on the C constant and smaller influence on the n constant, Figure 4.

As it was pointed above, an important issue regarding the propagation of fatigue cracks is the microstructure of the material. Moreover, the current trends in fatigue crack growth analysis are oriented towards the crack evolution from the early stages where the role of microstructure is a major one.

The role of microstructure on fatigue crack propagation has been highlighted in different studies published in the literature and therefore in this chapter is performed an analysis related to the effect of some particular microstructures on the fatigue crack growth rate.

Thus, in this analysis the influence of constituent phases and respectively the grain size and orientation on the fatigue crack growth rate are discussed.

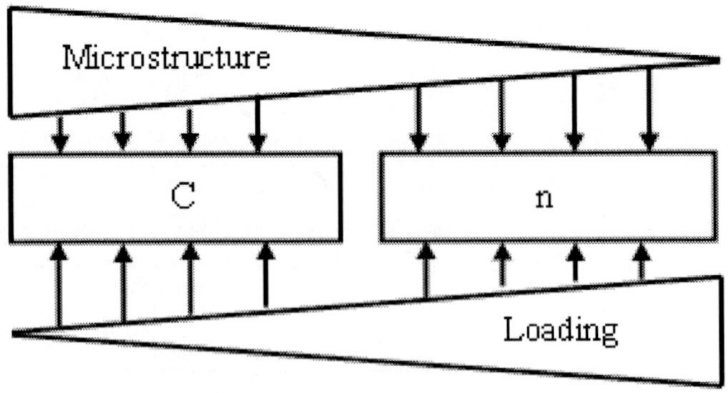

Figure 4. The effect of microstructure and loading conditions on propagation constants.

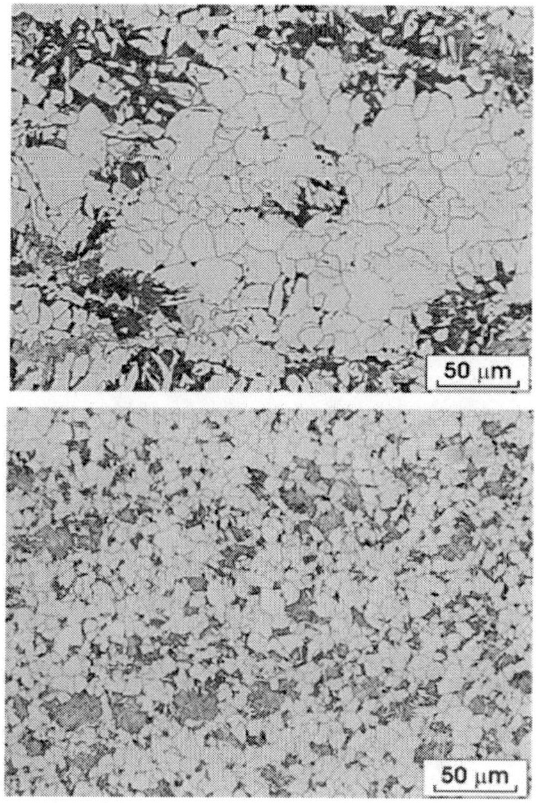

a)

b)

Figure 5. The microstructure of hot rolled low carbon ferritic-pearlitic steel plates: a) Steel N; b) Steel D. ([4] Korda et al.)

METHODS

2. The Influence of Second Phase on Fatigue Crack Growth Rate

2.1. Ferrite-Pearlite Microstructures

The experimental observations have shown that the volume fraction and morphology of second phase have significant effect on fatigue crack growth rate, [4-8]. Korda et al. [4] have studied the effect of pearlite morphology on the fatigue crack growth behavior of hot rolled low carbon steel plates. They studied two types of pearlite distribution in ferrite phase, with different volume fraction: the networked-like pearlite structure with 25.1% of volume fraction

(Steel N) and distributed pearlite structure with 20.3% of volume fraction (Steel D).

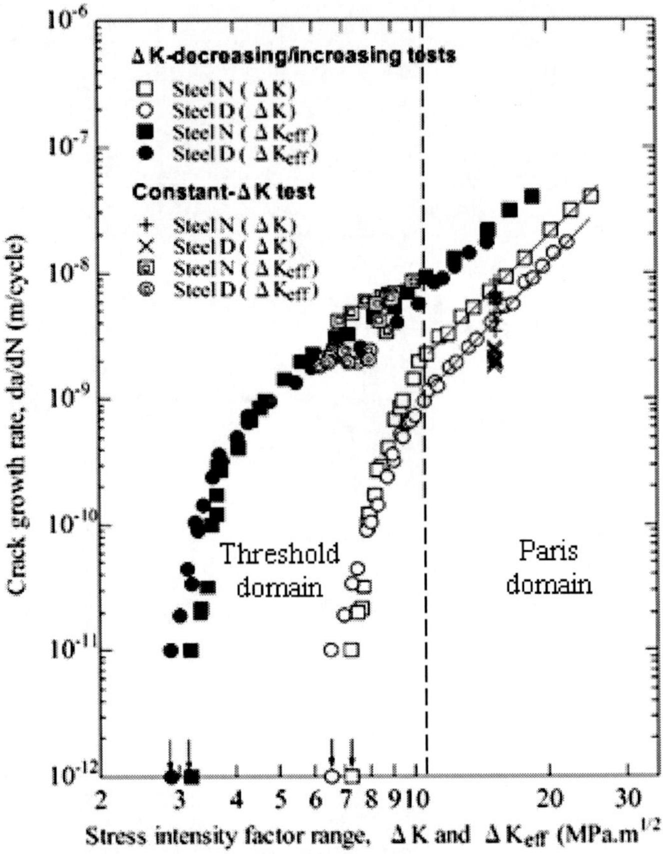

Figure 6. The fatigue crack growth rate variations function of ΔK and ΔK_{eff} for the two steels. ([4] Korda et al.).

In the Steel N, the threshold stress intensity factor range, ΔK_{th} value was 7.2 MPa$\sqrt{m}$. This value was higher compared to that in the Steel D, which was 6.5 MPa$\sqrt{m}$ for stress ratio 0.1. Their explanation for this difference between the threshold values was attributed to the grain size of the microstructure of both steels. The Steel N has coarser grain and lower yield strength than the Steel D. Also analyzing the fatigue crack propagation curves for the two steels there is a noticeable influence of the volume fraction and distribution mode of pearlite in ferrite on the fatigue crack growth rate and thus on the propagation constants, Figure 6. The Steel N with a greater volume fraction of pearlite has a lower fatigue crack propagation resistance than Steel D. For Steel N the propagation constant n is greater than that of Steel D instead the constant C is smaller than that of Steel D, Table 1. This means that in the Steel N the fatigue crack starts to grow at higher values of ΔK_{th}, the threshold domain is narrower and the fatigue crack growth rate is higher. In this case the effect of microstructure on crack propagation in Steel N is lower and the fatigue crack is mainly controlled by the applied load. In conjunction with pearlite morphology was observed that in the Steel D with distributed pearlite, the crack path was more tortuous compared to that in networked-like pearlite structure (Steel N). This tortuous path induced frequent crack interlocking.

Table 4. The propagation constants for Steels N and D

Material	R-ratio	C	n	ΔK_{th}
Steel N	0.1	$3.2 \cdot 10^{-14}$	4.202	7.2
Steel D	0.1	$1.739 \cdot 10^{-13}$	3.584	6.5

In another paper, [9] Korda et al. have studied the effect of banded pearlite in ferrite structure on fatigue crack propagation, Figure 7.

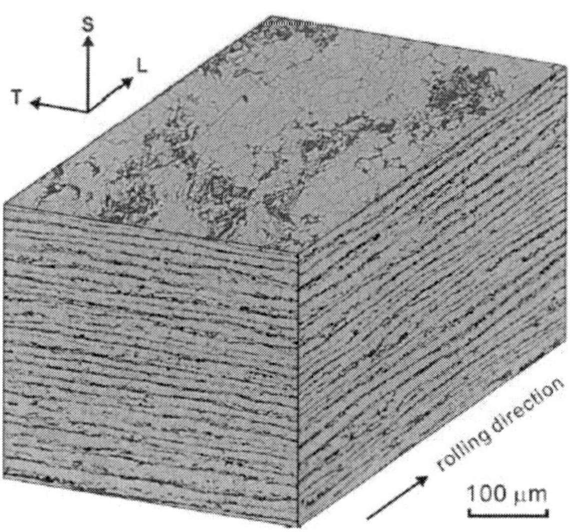

Figure 7. The pearlite-banded structure in the thickness direction (S). ([9] – Korda et al.).

Pearlite bands have been observed to arrest the propagation of cracks in fatigue tests. The pearlite bands provide barriers to crack growth and are considered to be responsible for the higher observed ΔK_{th} in the L-S orientation compared to the L-T orientation, Figure 7. In the pearlite-banded orientation, the frequent crack branching was observed along the crack path which reduced the local crack tip driving force for fatigue crack propagation. The crack closure in the pearlite banded orientation was more pronounced than in the non-banded orientation. This was done by the roughness-induced closure due to the large angle crack deflections at the ferrite-pearlite interface.

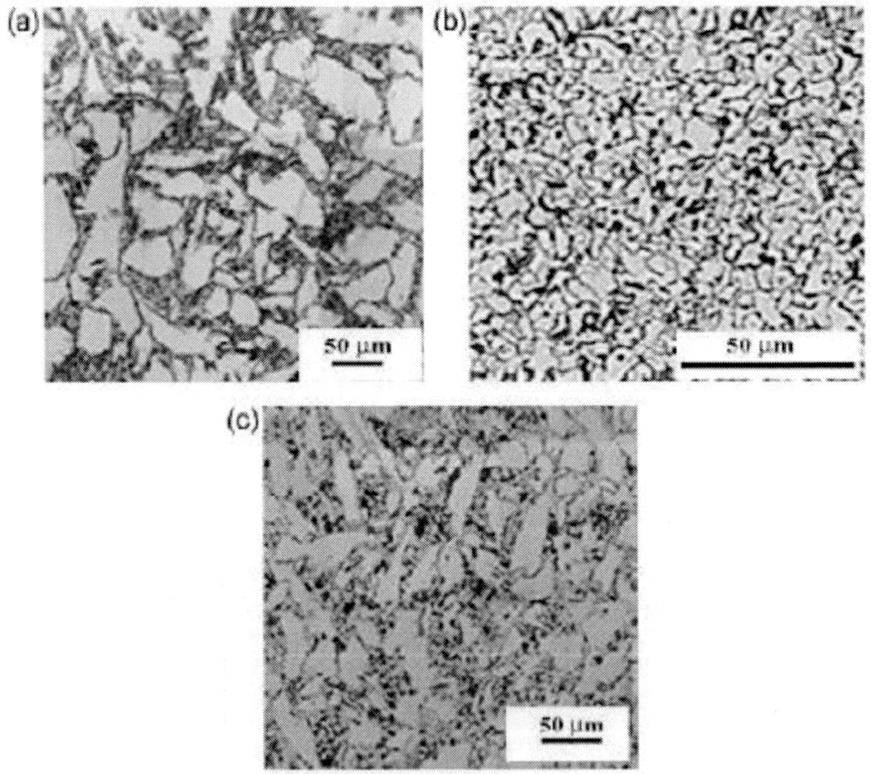

Figure 8. The three microstructures of ferrite-pearlite steel, ([10] – Mustapa and Mutoh): a) FP1 – ferrite grain size 50 µm, pearlite volume fraction 50%, size and spacing of pearlite particle 70 µm; b) FP2 - ferrite grain size 10 µm, pearlite volume fraction 50%, size of pearlite particle 5 µm, spacing of pearlite particle 12 µm; c) FP3 - ferrite grain size 65 µm, pearlite volume fraction 30%, size of pearlite particle 50 µm, spacing of pearlite particle 80 µm.

Mustapa and Mutoh [10] conducted an analysis on fatigue crack propagation in ferrite-pearlite steel with uniformly distributed pearlite particles. There were studied three microstructures with different sizes and spacing of pearlite particles, obtained by controlled heat treatment, Figure 8.

The microstructures FP1 and FP2 have the same volume fraction but different particle size and spacing. The microstructures FP1 and FP3 have different volume fractions of the pearlite phase and different pearlite particle size.

Based on the crack path observations, it seemed that more significant and frequently interlocking and branching of the crack were observed in FP1 and less frequently in FP3 and even less in FP2. Both interlocking and branching of the crack path are associated with crack tip stress shielding phenomena, Figure 9. Also the fatigue crack growth resistance of FP1 was the highest, whereas that of FP3 was the lowest.

The fatigue crack growth curves for the three materials were expressed by ΔK_{eff}, as shown in Figure 11. It was observed that the variation curves for FP1, FP2 and FP3 did not coincide with each other. This suggests that some mechanisms other than crack closure influence the fatigue crack growth resistance. As the interlocking and branching elements were often observed on crack path of all three materials, it was determined that the crack tip stress shielding was the corresponding mechanism, [10]. Moreover, if the fatigue crack growth rate for the three structures is correlated with $\Delta K_{eff,tip}$ (4), the results converge into a single curve, Figure 11, which supports the hypothesis that the crack tip stress shielding phenomena has an important role in fatigue crack propagation.

$$\Delta K_{eff,tip} = K_{tip} - K_{cl} \tag{4}$$

where $\Delta K_{eff,tip}$ is the effective crack tip stress intensity factor range, K_{tip}, Figure 10, is the crack tip stress intensity factor, K_{cl} is the closure stress intensity factor.

In addition to this was performed an analysis of the effect of pearlite morphology on fatigue crack propagation in relation to the propagation constants. Thus, in the case of FP1 and FP2 microstructures with the same volume fraction of pearlite (50%), the difference on the fatigue crack propagation behavior is given by the pearlite particle size and spacing. The distance between the pearlite particles has a major influence on crack deflection; instead the pearlite particle sizes has a major influence on the interlocking and branching elements and respectively crack closure. The fatigue crack growth rate is higher in the structure FP2 than the FP1 ($n_{FP2} > n_{FP1}$ and $C_{FP2} > C_{FP1}$). The large pearlite particle size in FP1 makes the fatigue crack to propagate at higher thresholds values than in FP2, respectively due to the contribution of interlocking and branching elements the propagation rate to be lower than in FP2. In the FP2 structure, the small pearlite particles size cause a reduction of interlocking and branching elements which favors a higher fatigue crack growth rate.

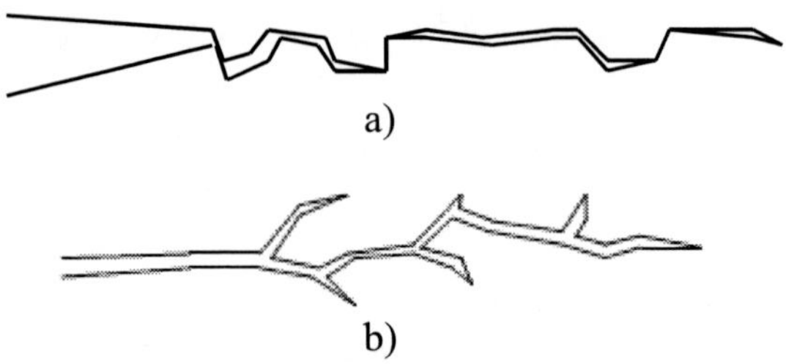

Figure 9. The crack tip stress shielding mechanisms: a) Interlocking; b) Branching.

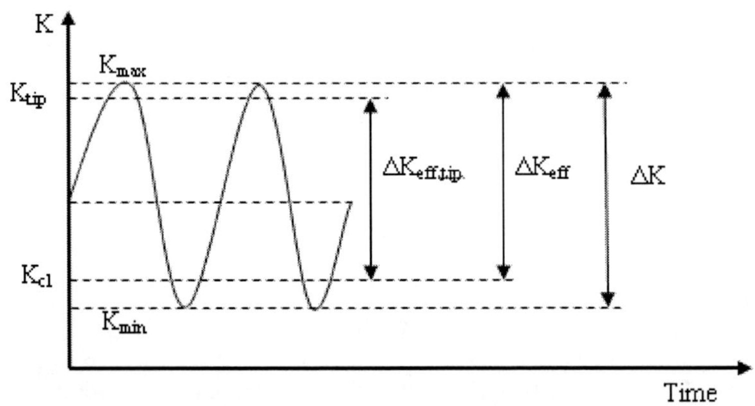

Figure 10. The determination of the crack tip effective stress intensity factor range.

In the FP3 structure the fatigue crack growth rate is higher than in FP1 on the Paris domain because the pearlite particles size is smaller, and interlocking and branching do not appear very often. Instead, on the threshold domain the fatigue crack growth rate in the structure FP1 is higher than in FP3 ($C_{FP1} > C_{FP3}$), still because the pearlite particle size is smaller in FP3 than the FP1. It is expected that the effective threshold stress intensity factor range corresponding to FP3 structure is lower than the corresponding value of FP1 ($\Delta K_{th,eff,FP3} < \Delta K_{th,eff,FP1}$).

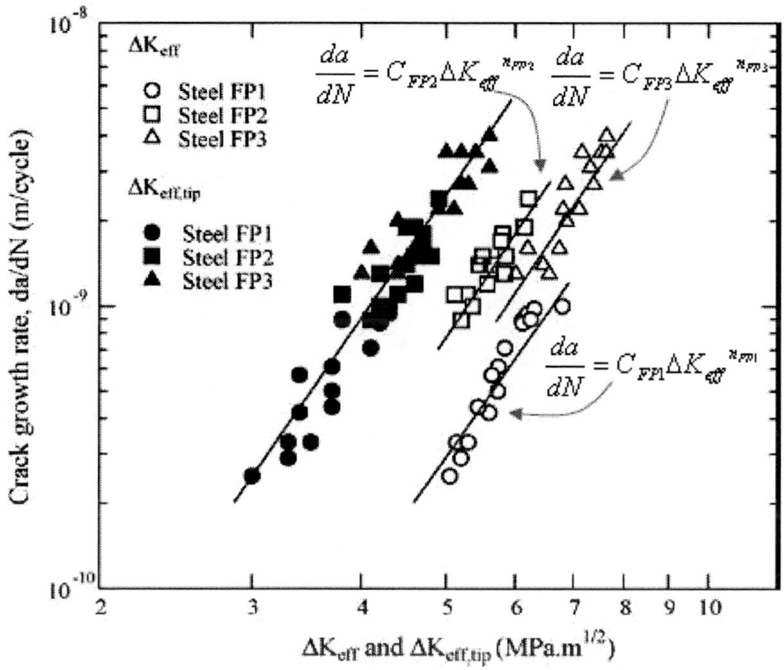

Figure 11. The fatigue crack growth curves in the three structures function of the effective stress intensity factor ranges ΔK_{eff} and $\Delta K_{eff,tip}$. ([10] Mustapa and Mutoh)

The constants C and n for the three structures were estimated based on experimental data of Figure 11 and are given in the Table 2.

Table 2. The propagation constants for the three analyzed structures

Material	R-ratio	C	n
FP1	0.1	$2.26 \cdot 10^{-13}$	4.475
FP2	0.1	$3.09 \cdot 10^{-13}$	4.78
FP3	0.1	$1.862 \cdot 10^{-13}$	4.796

Both the distribution mode and particle size of pearlite have a large influence on fatigue crack propagation in the ferrite-pearlite steels.

Based on this, in this analysis was performed an experimental study on fatigue crack growth rate in a low alloy steel plates with ferrite-pearlite microstructure. The material was analyzed in terms of chemical composition (Table 3), microstructure and mechanical properties (Table 4). The microstructural analysis indicated a low volume fraction of pearlite distributed in the ferrite phase, Figure 12.

Fatigue crack growth rate tests were conducted on CT samples, Figure 13, in the longitudinal and transverse direction of the steel plate, Figure 14, at a frequency of 3 Hz. The tests were carried out at room temperature and in laboratory air conditions, according to ASTM E 647 [11], while the crack length was monitored by the compliance method.

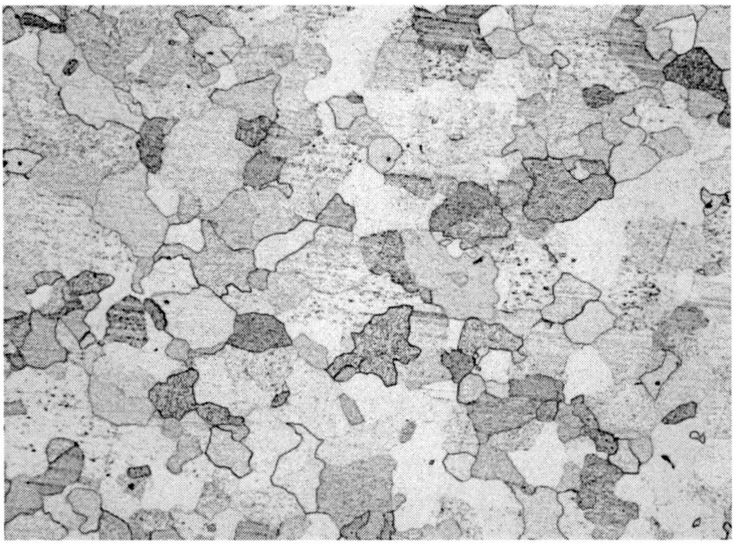

Figure 12. The microstructure of the analyzed steel.

Table 3. The chemical composition of the material [wt. %]

C	Si	Mn	P	S	Cu	Al	Cr	Mo	Ni	V
≤0.11	0.009	0.072	0.01	0.006	<0.003	0.023	0.005	0.026	0.013	0.005

Table 4. The mechanical properties of the analyzed steel

Material	Young's modulus [MPa]	Yield Strength [MPa]	Ultimate Tensile Strength [MPa]
Low-alloy steel	191000	175	287

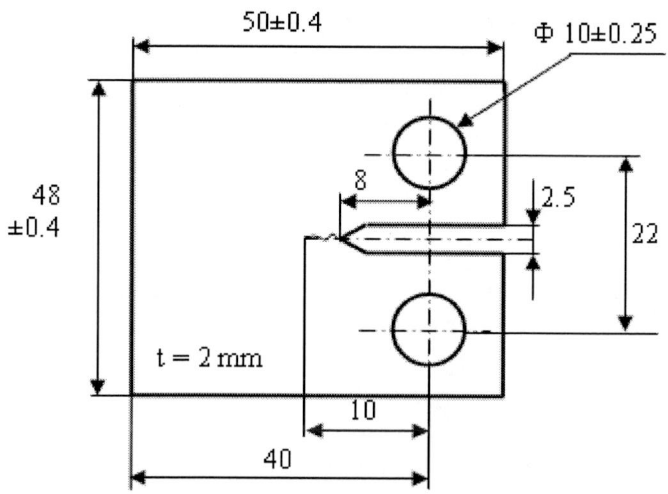

Figure 13. The CT specimen according to ASTM 647 standard.

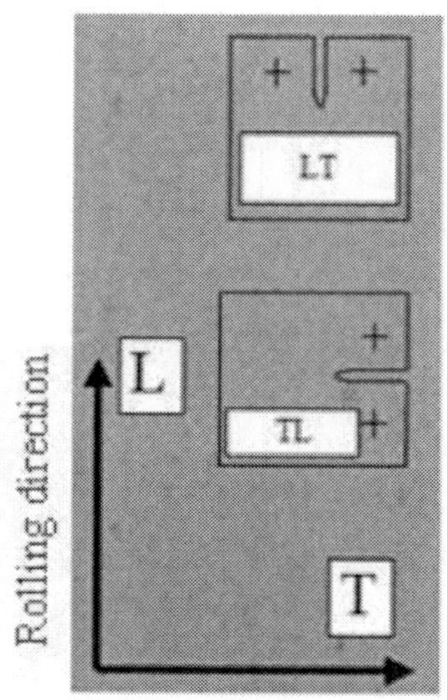

Figure 14. The directions of fatigue crack growth according to longitudinal and transverse direction of the steel plate.

The fatigue crack growth rate curves were determined in the longitudinal and transverse directions for cyclic loading with sinusoidal wave form and two stress ratios, 0.1 and respectively 0.5. The tests were conducted under K-increasing procedure as is described in ASTM E 647 standard. The crack closure load, P_{cl}, and corresponding closure stress intensity factor, K_{cl} were determined using the compliance offset technique. Based on these, the effective stress intensity factor range, ΔK_{eff}, was calculated.

$$\Delta K_{eff} = K_{max} - K_{cl} \quad (5)$$

Figures 16 and 17 show the fatigue crack growth rate curves for the two stress ratios and directions of the plate. The results show a significant R-ratio effect on fatigue crack growth rate in the threshold domain of the curve, corresponding to lower values of the applied stress intensity factor range, ΔK_{app}. As the applied stress intensity factor range increases to the stable propagation domain (Paris region) and the crack length increases, the R-ratio effect on the fatigue crack growth rate decreases. The threshold stress intensity factor range, ΔK_{th}, for the two stress ratios and crack growth directions were determined for the fatigue crack growth rate of $1 \cdot 10^{-10}$ m/cycle. Also, the crack propagation constants were defined and the results are given in Table 5.

Table 5. The propagation constants for analyzed ferrite-pearlite steel

Direction of the crack growth	R-ratio	C	n	ΔK_{th} [MPa√m]
LT (Longitudinal)	0.1	$1 \cdot 10^{-13}$	4.337	14
	0.5	$3 \cdot 10^{-12}$	3.365	9.89
TL (Transverse)	0.1	$4 \cdot 10^{-13}$	4.017	-
	0.5	$4 \cdot 10^{-12}$	3.223	10.33

In the case of fatigue crack growth in the longitudinal direction (coincident to the rolling direction of the steel plate) the propagation constant C is smaller than that corresponding to crack propagation perpendicular to the rolling direction (transverse direction, TL). Although, it could not accurately determine the threshold stress intensity factor range, ΔK_{th}, in the case of crack propagation in transverse direction, the variation curves indicates that ΔK_{th} corresponding to crack propagation in the longitudinal direction is smaller than that resulted for fatigue crack growth in the transverse direction. Instead, for the same stress ratio the propagation constant n for fatigue crack growth in the longitudinal direction is higher than that resulted from crack propagation in the transverse direction. The fatigue crack propagation behavior in the analyzed steel is affected by a combined effect of the rolling process and respectively the morphology of pearlite in the ferrite matrix.

The effect of the rolling process is more evident for R = 0.1 and causes a change in the propagation constants and the threshold value. Instead, for stress ratio 0.5 the effect of the rolling process is much lower, but there may be an influence of the microstructure on fatigue crack propagation in the two directions.

The effect of microstructure is reflected in the tortuous character of the crack path and respectively the high density of the interlocking and branches. It was observed that most of the crack branching occurs in the ferrite matrix due to misfit between the crystallographic characteristics of the grains, Figure 15.a. Instead, the areas with high density of pearlite are responsible for crack deflection. The fatigue crack tends to bypass the pearlite particles, Figure 15.b.

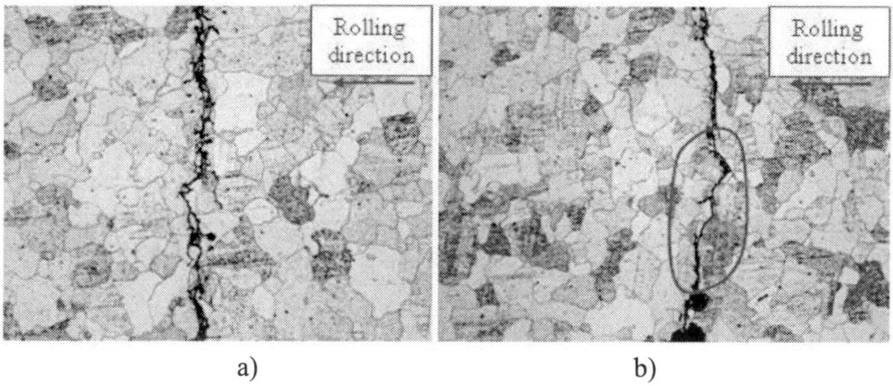

Figure 15. The fatigue crack growth path in transverse direction: a) interlocking and branching mechanisms; b) the deflection of crack path when encounters pearlite particle.

To highlight further the effect of microstructure, the fatigue crack growth rate data were correlated with the effective stress intensity factor range, Figure 18 and 19. The concept of crack closure and effective stress intensity factor range involve the removal of the stress ratio effect and thus the fatigue crack growth rates converge to a single curve. As shown in the figures below, in this case the data do not fit into a single curve, especially on the threshold domain of fatigue crack propagation in the longitudinal direction, Figure 18. This may indicate the presence of other mechanisms that influence the fatigue crack growth resistance.

The phenomenon through which the microstructure controls the fatigue crack propagation is represented by the crack tip stress shielding given by the interlocking and branching mechanisms. To quantify this effect on the fatigue crack growth rate for the two stress ratios and respectively the two directions of propagation, it is necessary to determine the crack tip stress intensity factor, K_{tip}, and respectively $\Delta K_{eff,tip}$.

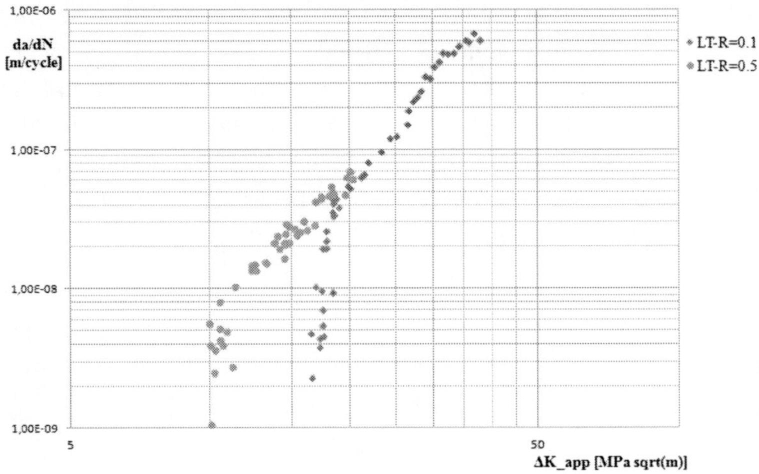

Figure 16. The fatigue crack growth rate function of applied stress intensity factor range, for crack growth in the longitudinal direction.

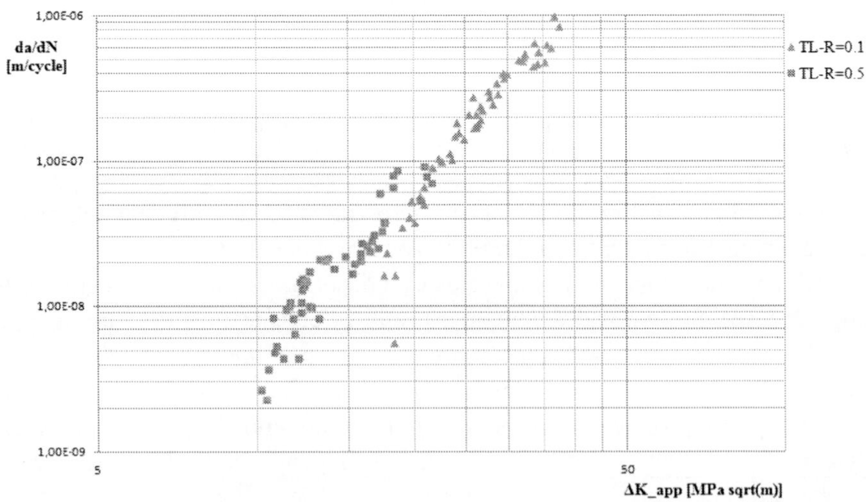

Figure 17. The fatigue crack growth rate function of applied stress intensity factor range, for crack growth in the transverse direction.

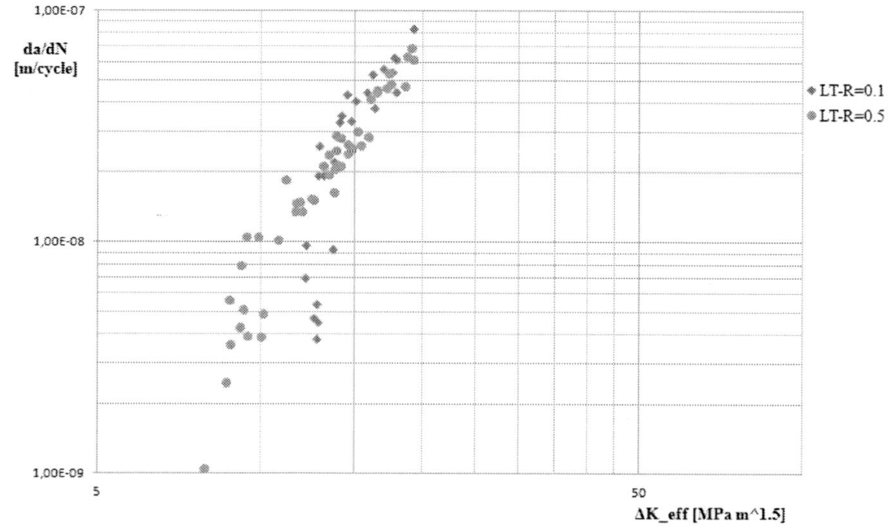

Figure 18. The fatigue crack growth rate function of effective stress intensity factor range, for crack growth in the longitudinal direction.

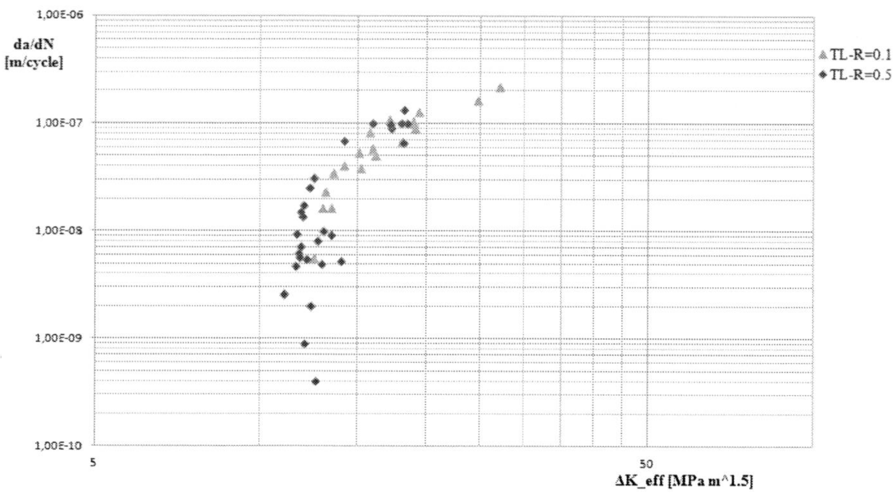

Figure 19. The fatigue crack growth rate function of applied stress intensity factor range, for crack growth in the transverse direction.

As defined in Figure 10, K_{tip} represents the real value of stress intensity factor at the crack tip. This value can be different from the maximum applied stress intensity factor, K_{max}, depending on the presence of interlocking and branching mechanisms. The maximum applied stress intensity factor is estimated based on an analytical solution that takes into account the specimen geometry, the maximum applied force and the crack length, [12].

Therefore, for each crack length was calculated the maximum stress intensity factor for the CT sample, $K_{max,t}$, based on the relationships below, and compared to the maximum stress intensity factor resulted from the tests, $K_{max,exp}$.

$$K_{\max,t} = \frac{F_{\max}}{t \cdot W^{1/2}} \cdot f(\alpha) \tag{6}$$

$$f(\alpha) = \frac{(2+\alpha)\cdot\left(0.886 + 4.64\cdot\alpha - 13.32\cdot\alpha^2 + 14.72\cdot\alpha^3 - 5.6\cdot\alpha^4\right)}{(1-\alpha)^{3/2}} \tag{7}$$

$$\alpha = \frac{a}{W} \tag{8}$$

where F_{max} is the maximum applied force, t is the thickness of the specimen, W is the width of the specimen, a is the crack length.

In Figure 20 is given the difference between the experimental and theoretical values of maximum stress intensity factor based on the crack length for all tests. It can be seen clearly that the crack tip stress shielding phenomenon has a large influence on fatigue crack propagation in transverse direction for both stress ratios. Instead, in the longitudinal direction the crack tip stress shielding phenomenon has a pronounced effect on fatigue crack propagation only for the stress ratio 0.1. Also, in the Figure 21 are shown the fatigue crack growth rates for the three situations where it is observed a pronounced effect of crack tip stress shielding, function of the effective crack tip stress intensity factor range, $\Delta K_{eff,tip}$ (4). It can see that the results converge to the same distribution band. The crack tip stress intensity factor, K_{tip}, was calculated by the following relationship:

$$K_{tip} = \frac{(\max COD - \min COD)\cdot K_{\max,\exp}}{\delta_0} \tag{9}$$

where max COD and min COD are the maximum and minimum crack opening displacements during one loading cycle, $K_{max,exp}$ is the maximum stress intensity factor resulted from fatigue crack propagation tests and δ_0 is the theoretical crack opening displacement [12]:

$$\delta_0 = \frac{P}{E \cdot t} \cdot V(\alpha) \tag{10}$$

$$V(\alpha) = \left(1 + \frac{0.25}{\alpha}\right)\left(\frac{1+\alpha}{1-\alpha}\right)^2 \left(1.6137 + 12.678\alpha - 14.231\alpha^2 - 16.61\alpha^3 + 35.05\alpha^4 - 14.231\alpha^5\right)$$

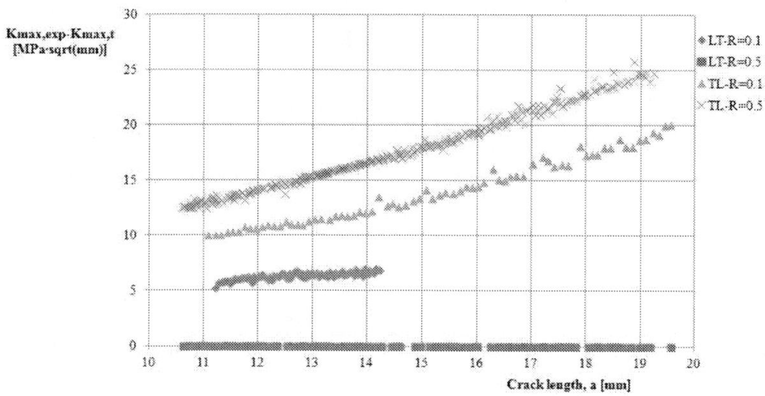

Figure 20. The difference between experimental and theoretical values of maximum stress intensity factors function of crack length.

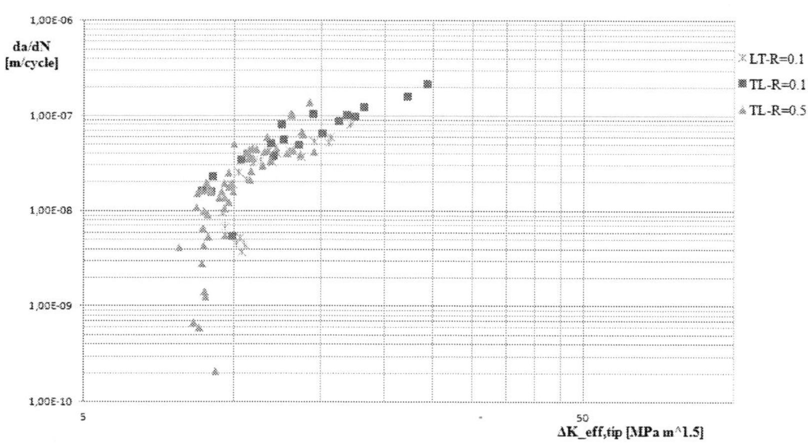

Figure 21. The variation curves of fatigue crack growth rate function of effective crack tip stress intensity factor range.

2.2. Ferrite-martensite Microstructures

The role of microstructure on fatigue crack growth in dual-phase steels has been the subject of many papers. Suzuki and McEvily [6] investigated fatigue crack growth behavior in two different dual-phase microstructures in an AISI 1018 steel. One was continuous ferrite matrix with encapsulated islands of martensite (FEM) and the other was continuous martensite with encapsulated ferrite islands (MEF). They found that the MEF microstructure had a higher threshold value and fatigue crack growth resistance than that in FEM microstructure. Minakawa et al. [13] and Ramage et al. [14] found that the connectivity of hard phase martensite had significant effect on fatigue crack growth rate and threshold value. A high degree of connectivity of martensite can lead to a high threshold value.

Wang and Yan [15] have studied the fatigue crack propagation behavior in three dual-phase structures obtained by well controlled heat treatments. Also, it was used for comparison an as-rolled dual-steel phase. The three structures consisted of: acicular martensite in ferrite matrix (A); netted martensite along ferrite boundaries (B) and granular martensite islands surrounded by ferrite matrix (C). Also, the structures A, B and C had the same grain size of about 50 μm, much larger than that of the as-rolled steel (10 μm). All structures were tested at constant amplitude loading with stress ratio 0. In Table 6 are listed the threshold values, ΔK_{th}, in different cases. The results showed that the morphology and respectively the arrangement of martensite have a significant effect on the fatigue crack growth rate and the threshold stress intensity factors range. For the structures A and B the threshold values are higher and the fatigue crack growth rate in the threshold domain is lower than for the structure C. This means that the propagation constants for the structures A and B will be lower than those corresponding to structure C.

Table 6. Threshold values of dual-phase steels with different morphologies of martensite, [15]

	Material			
	A	**B**	**C**	**as-rolled**
ΔK_{th} [MPa$\sqrt{m}$]	16	17	12.8	8.1

The microstructure with continuous (or netted) martensite can produce the crack closure. Such a microstructure based on martensite will shield the crack by acquiring a great part of the applied load and thus reduces the driving force at the crack tip. The martensite being a constituent with high strength can take from the magnitude of the applied load. In the continuous ferrite microstructure (with granular martensite) the shielding effect is given by the crack branching. The threshold values for all structures are much higher than that corresponding to as-rolled steel. The threshold stress intensity factor range increases with increasing the grain size and decreasing the yield strength.

The effect of martensite content on the fatigue crack propagation has technological importance in choosing the optimal conditions for heat treatment of dual-phase steel. In the References [16-18] is studied the effect of martensite content with volume fraction from 5.1% to 82% and similar morphology, on the materials behavior. The increasing of volume fraction of martensite increases the yield and ultimate tensile strength and decreases the elongation. In terms of fatigue crack propagation, it was observed that up to V_m = 35% the fatigue crack growth rate decreases and the threshold stress intensity factor range increases with increasing the volume fraction of martensite. The threshold value decreases as the volume fraction of martensite increases over 35%.

The variation of the threshold stress intensity factor range with the martensite content does not have linear relationship as reported for ferrite/pearlite or bainite/martensite microstructures. It was also observed that at dual-phase steels with volume fraction of martensite less than 30-50% occurs cyclic hardening, while for content exceeding 50% cyclic softening occurs.

2.3. Ferrite-Bainite and Bainite-Martensite Microstructures

The bainite is another hard constituent that may have a significant influence on the fatigue crack propagation in dual-phase steels. Bainitic steel has received considerable interest due to its better combination of strength and ductility, as compared to full untempered martensite, which is achieved by controlling cooling conditions, [19]. The microstructures and mechanical properties of bainitic steels vary, in relation to chemical composition and cooling conditions. In martensite/bainite mixed structures the shape and distribution of the second phase bainite have a significant effect on the K_{IC}. The fracture toughness values of ferrite-bainite-martensite were reported to be inversely proportional to the yield strength and were attributed to the ductile nature of microstructure, [20].

Kim T. and Kim H. [19] conducted an experimental study on lower bainitic medium-carbon steel (SCM 435). Fatigue crack growth rate tests were performed on two microstructures obtained by controlled heat treatments at constant amplitude loading with stress ratio 0.1. The two microstructures analyzed were tempered martensite with 19.2% lower bainite (LB-M) and fully lower bainite. The results indicated that the fatigue crack growth rate in the fully lower bainite structure is lower than that in LB-M. The lower bainite appears in acicular form which partitions prior austenite grains. This makes the crack starts to propagate at lower stress intensity factor range, but in the same time the bainitic microstructure with hard carbide particles densely distributed in the ferrite matrix determine the crack propagation in a larger threshold domain with higher fatigue crack growth rate. Also, on the threshold domain the fatigue crack propagation is controlled by the branching and interlocking mechanisms. Moreover, the fatigue crack propagation behavior on the threshold domain generates an increase of propagation constant C. Once the fatigue crack reaches the Paris domain, the combination of strength and ductility of bainite causes a reduction of propagation constant n, through the contribution of the plastic deformation given by bainite.

Table 7. The propagation constants from Paris 'equation, for stress ratio 0.1, [19]

Material	Propagation constants	
	C	n
LB-M	$8.5\cdot 10^{-12}$	3.01
Fully lower bainite	$1.1\cdot 10^{-11}$	2.78

Comparing two dual-phase structures (ferrite-pearlite, F-P and ferrite-bainite, F-B) Guan and Yu [21] showed that bainite described as a hard phase distributed in ferrite matrix introduces a shielding effect on fatigue crack propagation by branching and interlocking. The crack path in F-P steel was more tortuous compared with that in F-B. In the Paris domain, fatigue crack growth rate of F-B steel was lower than that in F-P.

Xu et al. [22] investigated the fatigue crack propagation behavior in three types of microstructure steels, for constant amplitude loadings with different stress ratios. The three microstructures were ferrite-pearlite (FP: 86%-14%), ferrite-bainite-martensite (FMB: 8%-17%-75%) and tempered martensite (TM: 2% bainite-98% martensite). The fatigue crack growth rate tests on CT specimens showed that for stress ratio 0.1 in Paris domain the fatigue crack growth rate of FP steel was higher than in the other two steels (FBM and TM). In Table 8 are given the propagation constants for the three materials, for stress ratio 0.1 and also the corresponding threshold values. In the threshold domain the crack propagation path of the FP steel presents a more tortuous appearance compared with the other steel structures. The crack propagation path of the FBM and TM steels are flatter. The tortuous crack propagation path of the FP steel can be attributed to branching and deflection that occurs when the crack propagating in the soft ferrite encounters the harder pearlite.

Table 8. The propagation constants from Paris 'equation, for stress ratio 0.1 [22]

Material	Propagation constants		ΔK_{th}
	C	n	[MPa√m]
FP	$6.3 \cdot 10^{-13}$	3.496	9.03
FBM	$2.91 \cdot 10^{-12}$	3.116	7.97
TM	$1.2 \cdot 10^{-11}$	2.725	7.08

The high values of propagation constant C for TM and FBM structures compared to FP structure are due to differences between grain size and respectively the morphology and properties of hard constituents (martensite, bainite and pearlite). Also, on the Paris domain the difference between the propagation constants n for the three structures is attributed to the strength of martensite and ductility of bainite that make the structures FBM and TM to have a higher fatigue crack growth resistance than the FP structure.

3. The Influence of Grain Size and Orientation on Fatigue Crack Growth Rate

The grain size affects the tensile properties of materials which exhibit either cross-slip or difficult cross-slip. Thus, only the latter materials show an influence of grain size on long fatigue lives [23]. For long fatigue lives, grain size and slip mode were found to affect fatigue cracking only during threshold domain. The effect of grain size on the fatigue crack growth rate in threshold domain of AISI 1018 steel was studied by Gray et al. [24]. Grain sizes were equal to 100 μm for coarse-grain and 20 μm for fine-grain structure, while yield strength levels of both structures were equivalent. Fatigue crack growth rate tests were conducted for stress ratios 0.1 and 0.7. The findings were that at stress ratio 0.1 the fatigue crack growth rates in both threshold and Paris domains of the coarse-grain structure are lower than that of the fine-grain structure. At a high R ratio of 0.7 the fatigue crack growth rate of both coarse and fine-grain structures are essentially comparable only in the threshold domain. L.A. James [25] found that higher fatigue crack growth rates are associated with fine-grain structure and lower growth rates with larger-grain structure. Also, in the threshold domain Taira et al. [26] showed that the fatigue crack growth rate is lower in the material with larger grain size than that in smaller-grain size material when compared at the same stress intensity factor range. There were studied three grain sizes (7.8, 20.5 and 55 μm) and observed that at the same R ratio the value of ΔK_{th} increases linearly with the square root of the grain size, $\sqrt{d}$ (d is the grain size). Also, an experimental equation was proposed for the threshold value of ΔK_{eff} function of $\sqrt{d}$:

$$\Delta K_{eff,th} = 0.19 + 196 \cdot \sqrt{d} \tag{11}$$

At the grain level the fatigue crack propagation is characterized by intragranular and intergranular fractures. These fracture types depend on the structural constituents (as shown above) but also by the grain boundaries and orientation.

The intragranular fracture occurs due to plastic deformation developed at the grain level by slipping of crystallographic planes. The slip always occurs on a particular set of crystallographic planes, known as slip planes. Inside of these planes, the slip occurs along a consistent set of directions known as slip directions. The combination of slip plane and slip direction together makes a slip system.

The slip occurs when the shear stress acting in the slip direction on the slip plane reaches some critical value. The critical shear stress is related to the stress required to move dislocations across the slip plane. It was found that the critical value of shear stress at which slip occurs in a given material with specified dislocation density and purity, is a constant known as the critical resolved shear stress, τ_c. This is also known as Schmid's law, [27].

The tensile stress at which the material starts to slip is the yield strength and the relation with critical resolved shear stress is given by the following equation:

$$\tau_C = \sigma_y \cdot \cos\varphi \cdot \cos\lambda \tag{12}$$

where σ_y is the yield strength, φ is the angle between the loading direction and normal to slip plan and λ is the angle between loading direction and slip direction, Figure 22.

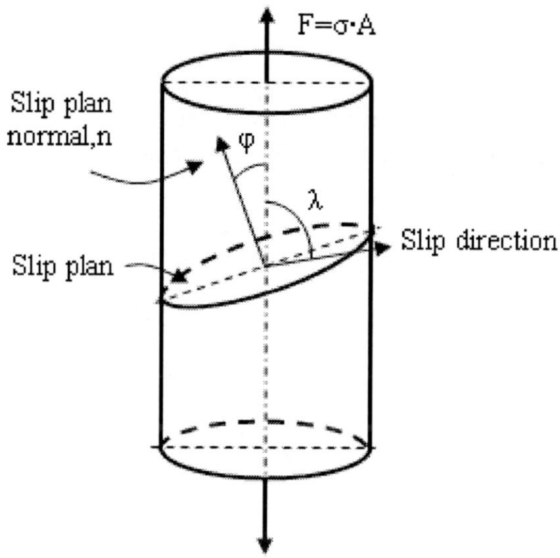

Figure 22. The slip characteristics of crystallographic plan.

The product cos $\varphi \cdot$ cos λ is defined as Schmid factor, (M) and it was found that the crystal begins to plastically deform in the slip system with greatest Schimd factor, named primary system.

The intergranular fractures are characterized by the detachments of the grains due to either grain orientation or their different nature. Chen et al. [28] have shown that the fatigue crack growth rate decreases significantly when the crack tip that crosses a grain meets its boundaries. The movement of dislocation will be impeded by the grain boundaries, which contributes to the evident decrease of fatigue crack growth rate. The fatigue crack growth rate increases rapidly once the crack has traversed the grain boundaries.

Also, it is generally known that during loading and deformation of the grains, stress concentration occurs at boundaries due to accumulation of dislocations. These occur especially for high-angle grain boundaries where the deformation of neighboring grains can be hardly harmonized and dislocations are difficult to cross. When the high-angle grain boundaries are arranged nearly normal to loading direction they will experience much higher tensile stress than those parallel to loading direction (red arrows in Figure 23). Therefore, during fatigue loading that grain boundaries parallel to loading direction are more susceptible to crack initiation.

Grains with low Schimd factor are known to be harder to activate slip during deformation and the fatigue crack extension depends on dislocation slipping within the plastic zone of the crack tip. Thus, fatigue crack is more likely to propagate into a softer grain when it confronts grains with different Schmid factors. Moreover, it was found that Schimd factor distribution has no obvious effect on the intergranular crack initiations. The fatigue crack can initiate at grain boundaries between either soft grains or hard grains, [28].

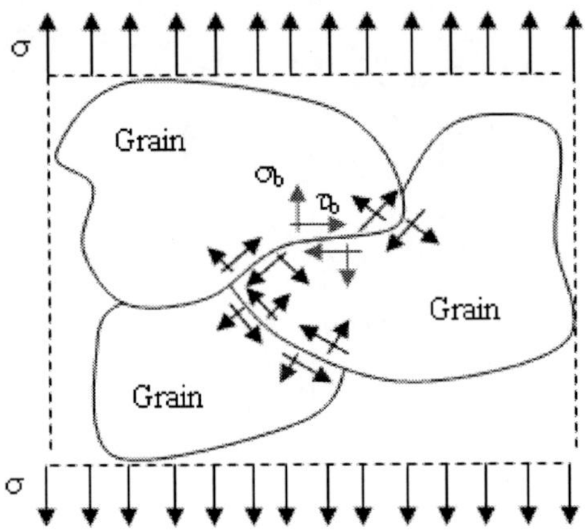

Figure 23. The stress concentration at grain boundaries: σ is the applied stress and σ_b and τ_b are the normal and shear stresses at grain boundaries.

Chakravarthula and Qiao [29] conducted an experimental study related to fatigue crack growth rate in a Fe-2% Si alloy where it was analyzed the role of grain boundaries. They found that the crack front must be necessarily branched geometrically to overcome a high-angle grain boundary. The crack penetrates across the grain boundary at a number of break-through points (BTP) that are distributed along the front at the distance D. When the width, w, extends over a certain value from the persistent grain boundary island (PGBI), the grain boundary is overcome and the crack continues to propagate into adjacent grain, Figure 24.

Also, based on this model, there are two options available for fatigue crack propagation through grain boundaries. The first possible mode of PGBI separation is the coplanar fatigue crack growth along the boundary, Figure 25.a. In this option a mode II fatigue crack growth is triggered along the grain boundary. When the length of mode II intergranular crack reaches the critical value, c_{cr}, the height difference between the adjacent crack planes is reduced to h_{cr} and the grain boundary resistance can be overcome at the maximum stress intensity factor, K_{max}, [29].

$$K_{max} = \sqrt{\frac{Ekw}{2(1-\nu^2)}} \cdot \tan\varphi \cdot (1-2\tilde{c}) \tag{13}$$

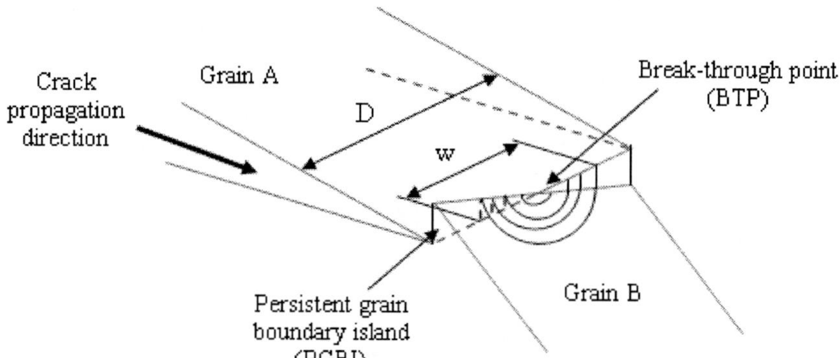

Figure 24. The geometrically crack front branching across a high-angle grain boundary [29].

where E is the Young's modulus, k is the effective shear strength of the grain boundary, w is the width of fatigue crack, ν is the Poisson's ratio, φ is the twist crystallographic misorientation and $\tilde{c} = c_{cr} / h_0$.

If h_0 is estimated as w · tan φ, [29], then from Equation (13), it can be written as:

$$\frac{2c_{cr}}{h_0} = 1 - K_{max} \cdot \frac{\cot\varphi}{\sqrt{\frac{Ekw}{2(1-\nu^2)}}} \tag{14}$$

The second mode to cross the grain boundary by the fatigue crack is through the separation of PGBI by an abrupt fracture process, Figure 25.b. During the cyclic loading, the fatigue crack front penetrates the grain boundary at the break-through points. As the penetration depth becomes larger, more grain boundary areas are separated, leading to a lower rate of fracture work increment. Due to the limitation of the grain boundary geometry, there is a critical penetration depth, a_{cr}, at which the crack growth driving force becomes larger than the rate of the work to separation PGBI and the crack overcome the grain boundary, [29]. The critical condition is given by:

$$\frac{K_{\max}}{K_{SC}} = \left[1 + 0.12k_0 \left(\frac{\sin^2\varphi \cdot \cos^2\varphi \cdot \cos\psi}{\sin\varphi + \cos\varphi}\right)^{1.05}\right] \cdot \sqrt{\frac{\sin\varphi + \cos\varphi}{\cos\psi}} \quad (15)$$

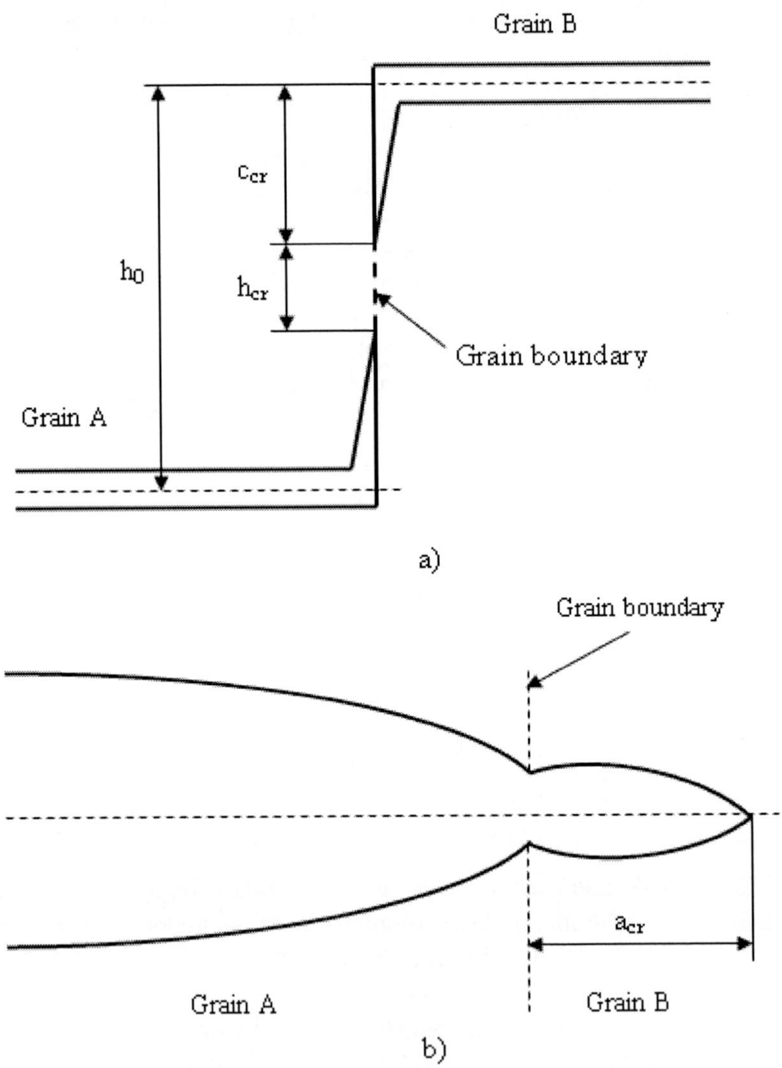

Figure 25. Two possibilities for a fatigue crack growth through a high-angle grain boundary: a) coplanar fatigue crack growth along the grain boundary; b) abrupt separation of the persistent grain boundary islands [29].

where K_{SC} is the fracture toughness of a single crystal, ψ is the tilt crystallographic misorientation, and:

$$k_0 = \left[\left(\frac{a_0}{w} \right)^{\beta} \cdot \left(\frac{\alpha^3}{8} \right) \cdot \frac{k}{\mu} \cdot \frac{kw}{G_{SC}} \right]^{1.05} \tag{16}$$

with a_0 being the crack length, α and β the geometry factors of the penetration front, μ the shear modulus and G_{SC} the fracture resistance of a single crystal. The critical penetration depth is given by:

$$\frac{a_{cr}}{w} = \left[\frac{2(1-\nu^2)K_{\max}^2}{E(a_0 + a_{cr})\beta k^*} \cdot \frac{1}{(\sin\varphi \cdot \cos\varphi)^2} \right]^{\frac{1}{\beta-1}} \tag{17}$$

$$k^* = \frac{\alpha^3 k^2}{8\mu} \tag{18}$$

More recently, Wei et al. [30] studied the influence of grain structure and crystallographic orientation on fatigue crack growth rate in 7050 alloy thick plate. They observed that the intragranular and intergranular fatigue crack, crack deflection and bifurcation depend on local microstructural characteristics, crystallographic orientation and geometry of grain with respect to the loading direction. The difference between the twist angles of the slip planes from two adjacent grains has a significant role in extending the crack from a grain to the other. Thus it was observed that a large difference between the twist angles makes the crack to not extend into the adjacent grain, but instead to propagate along the grain boundary. On the other hand, in the grain both the difference between the twist and tilt angles of the previous crack plane and possible slip plane are responsible for crack deflection.

The tilt and twist angles have been defined by the Eq. (19) and (20) and are shown in Figure 26.

$$\theta_i = \arccos([slip]_i \times [surface]_i \bullet [loading]_i) \tag{19}$$

$$\psi_i = \arccos\left([slip]_i \times \left[grain \quad boundary \right]_i \bullet [loading]_i \right) \quad (20)$$

where θ_i is the tilt angle of crack plane in grain i, ψ_i is the twist angle of crack plane in grain [slip]$_i$ is the slip plane of grain i, [surface]$_i$ is the normal direction of sample surface, [grain boundary]$_i$ is normal direction of the grain boundary and [loading]$_i$ is the loading direction.

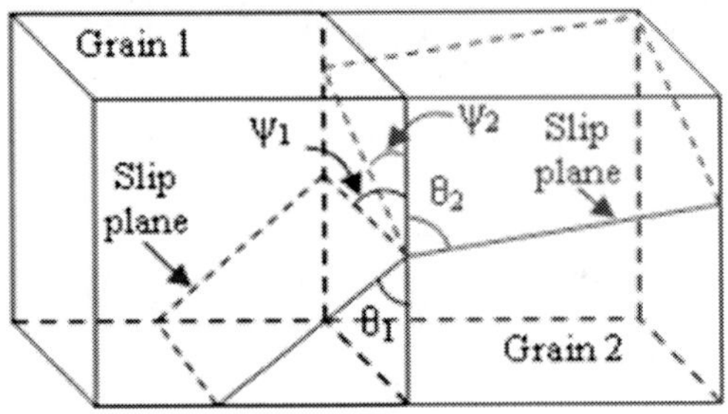

Figure 26. The crystallographic characteristics with effect on fatigue crack growth across the grain boundary [30].

4. Discussion and Conclusion

The presence of hard constituents in microstructure can have a significant influence on the fatigue crack propagation. It was shown that the presence of distributed pearlite in ferrite matrix increases the fatigue crack propagation resistance compared with networked-like pearlite. Distributed hard pearlite in soft ferrite matrix determines a retardation effect of the fatigue crack propagation by introducing the crack tip stress shielding phenomenon. This phenomenon is reflected on fatigue crack propagation by the branching and interlocking mechanisms. Also, on the Paris domain, the crack tip stress shielding causes a decrease of propagation constant n. On the other hand, if the pearlite is distributed in the ferrite matrix in the form of large size particles (see the steel FP1) the effect on the fatigue crack propagation is evident for the threshold domain. In this case the fatigue crack starts to propagate at higher threshold values, the fatigue crack growth rate is higher, the threshold domain is narrow, resulting in an increase on the propagation constant C. The effect of pearlite morphology on fatigue crack propagation is illustrated in Figure 27.

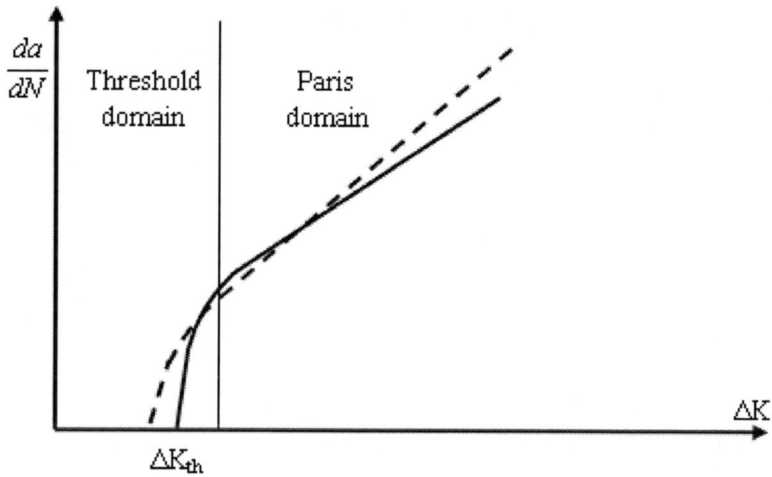

Figure 27. The effect of pearlite morphology on the fatigue crack growth behavior.

In the case of fatigue crack propagation in hot rolled ferrite-pearlite steels, in addition to pearlite morphology the crack growth direction may influence the fatigue crack propagation behavior. Thus, in the case of fatigue crack propagation normal to rolling direction the crack tip stress shielding effect is higher. Also, the effect of this phenomenon has been preserved with pronounced character for both stress ratios considered. Instead, in the case of fatigue crack propagation on rolling direction, the crack tip stress shielding effect has decreased considerably for stress ratio 0.5. The fatigue crack is retained by the particles elongated on the rolling direction which causes more branching and interlocking. The fatigue crack growth rate is higher on the rolling direction compared to normal direction.

The distribution of acicular bainite in ferrite makes the fatigue crack to propagate at lower threshold values than in the case of ferrite-pearlite structures. Instead, on the Paris domain the strength and ductility of the bainite lead to increase the fatigue crack propagation resistance by decreasing the propagation constant n.

Although the microstructural grain size affects the fatigue crack growth rate only on the threshold domain, their structure may have a dominant role during the entire crack propagation. Thus, it was observed that the fatigue crack propagates through the grains with high Schimd factor, which allows easy activation of slip planes, and respectively tends to bypass those grains with low Schmid factor. The analytical models previously presented and developed based on experimental observations highlight the fact that in addition the structure, also the crystallographic characteristics and grain boundaries orientation influence the fatigue crack propagation by introducing the crack tip stress shielding effects. The fatigue crack can be retarded by the grain boundary after its tip passed the limiting grain surface due to a high twist angle of the grain boundary in its plane (Figure 24). At the same time the tilt and twist angles of the crack plane which crossed a grain and the corresponding tilt and twist angles of the possible slip planes from the adjacent grain can determine the crack deflection. Also, the crack deflection during transgranular propagation can be attributed to the cooperation between slip systems and loading stress.

In terms of fatigue behavior of materials in relation to the microstructure, one of the current trends of research consists in the ability to predict the initiation and propagation of fatigue cracks in terms of a given microstructure.

Both the trends and research directions on monitoring the evolution of fatigue crack from the early stages of initiation and then the propagation, implies the need for a change of approach in terms of predictive analyzes. One of the main changes in approach will consist in the defining of the material element, which besides their macro elastic characteristics (eg, E and ν) should be considered also the microstructural parameters (constituent phases, grain size and orientation). This is supported on the one side by the analytical models already developed and on the other hand the technological capability of reconstructing the microstructure, [31, 32 and 33].

The role of microstructure on fatigue crack propagation is an extensively studied topic, but on how much it is analyzed throughout that much open new discussion that require the development of new studies. It is generally understood that the material microstructure influence the properties and their behavior but how are influenced is still an open discussion. This is due to the heterogeneity of materials at micro level and respectively diversity of microstructures that can be obtained through various technologies.

The experimental observations discussed in this chapter may constitute necessary elements for the development of analytical models and methodologies, but equally may constitute basic elements in establishing the technological conditions for materials development.

REFERENCES

[1] *Aircraft Structural Integrity Requirements,* MIL-STD 1530, U.S. Air Force, 1996.

[2] *NASGRO Fracture Mechanics and Fatigue Crack Growth Analysis Software*, Version 4.02, NASA-JSC and SwRI, 2002.

[3] J.C. Newman, *A crack opening stress equation for fatigue crack growth*, International Journal of Fracture, 24: R131-R135, 1984.

[4] A.A. Korda, Y. Miyashita, Y. Mutoh, T. Sadasue, *Fatigue crack growth behavior in ferritic-pearlitic steels with networked and distributed pearlite structures*, International Journal of Fatigue, 29: 1140-1148, 2007.

[5] J.A. Wasynczuk, R.O. Ritchie, G. Thomas, *Effects of microstructure on fatigue crack growth in duplex ferrite-martensite steels*, Mater. Sci. Eng, 79-92, 1984.

[6] H. Suzuki, A.J. McEvily, *Microstructural effects on fatigue crack growth in a low carbon steel*, Metall. Trans. A, 10A: 475-81, 1979.

[7] R.M. Ramage, K.V. Jata, G.J. Shiflet, E.A. Starke Jr, *The effect of phase continuity on the fatigue and crack closure behavior of a dual-phase steel*, Metall. Trans. A, 18A: 1291-8, 1987.

[8] T.M. Hashimoto, M.S. Pereira, *Fatigue life studies in carbon dual-phase steels*, Int. J. Fatigue, 18: 529-33, 1996

[9] A.A. Korda, Y. Mutoh, Y. Miyashita, T. Sadasue, S.L. Mannan, *In situ observation of fatigue crack retardation in banded ferrite-pearlite microstructure due to crack branching*, Scripta Materialia 54: 1835-1840, 2006.

[10] M.S. Mustapa, Y. Mutoh, *Effects of size and spacing of uniformly distributed pearlite particles on fatigue crack growth behavior of ferrite-pearlite steels*, Materials Science and Engineering A, 527: 2592-2597, 2010.

[11] ASTM E647, *Standard test method for measurement of fatigue crack growth rates.*

[12] Y. Murakami, *Stress Intensity Factors Handbook*, vol. 1, Pergamon Press, Oxford, 1987.

[13] K. Minakawa, Y. Matsuo, A.J. McEvily, *Influence of a duplex microstructure in steels on fatigue crack growth in the near-threshold region*, Metall Trans A, 13A(3): 439-45, 1982.

[14] R.M. Ramage, K.V. Jata, G.J. Shiflet, E.A. Starke Jr., *The effect of phase continuity on the fatigue and crack closure behavior of a dual-phase steel*, Metall Trans A, 18(A):1291-8, 1987.

[15] Z.G. Wang, M.G. Yan, *Fatigue crack propagation in advanced structural materials*, Handbook of Fatigue Crack Propagation, A. Carpinteri editor, Elsevier, vol. 1, p. 323, 1994.

[16] D.L. Chen, *Near-threshold fatigue crack growth in dual-phase steels*, PhD thesis, Institute of Metal Research, Chinese Academy of Sciences, China, 1989.

[17] Z.G. Wang, D.L. Chen, X.X. Jiang, C.H. Shih, B. Weiss, R. Stickler, Proc. Fatigue '90 (Edited by H. Kitagawa and T. Tanaka), vol. 3, p. 1363, Birmingham, Materials and Component Engineering Publications, 1990.

[18] Z.G. Wang, D.L. Chen, X.X. Jiang, S.H. Ai, C.H. Shih, Proc. Int. Conf. on Basic Mechanisms in Fatigue of Metals (Edited by P. Lukas and J. Polak), p. 351, Brno, 1988.

[19] T.Y. Kim, H.K. Kim, Mixed-mode fatigue crack growth behavior of fully lower bainite steel, Materials Science & Engineering A, 580: 322-329, 2013.

[20] Y. Tomita, *Effect of microstructure on plane-strain fracture toughness of AISI 4340 steel*, Metall. Trans. A, 19A:2513-2521, 1988.

[21] M. Guan, H. Yu, *Fatigue crack growth behaviors in hot-rolled low carbon steels: A comparison between ferrite-pearlite and ferrite-bainite microstructures*, Materials Science & Engineering A, 559: 875-881, 2013.

[22] T. Xu, Y. Feng, S. Song, D. Wang, *Fatigue crack propagation behavior of steels with different microstructures*, Materials Science and Engineering A, 551: 110-115, 2012.

[23] A.W. Thompson, R.J. Bucci, *The dependence of fatigue crack growth rate on grain size*, Metall. Transactions, vol. 4: 1173-1175, 1973.

[24] G.T. Gray, J.C. Williams, A.W. Thompson, *Roughness-induced crack closure: an explanation for microstructurally sensitive fatigue crack growth*, Metall. Trans. A, vol. 14(2): 421-433, 1983;

[25] L.A. James, *The effect of grain size upon the fatigue-crack propagation behavior of alloy 718 under hold-time cycling at elevated temperature*, Engineering Fracture Mechanics, 25(3):305-314, 1986.

[26] S. Taira, K. Tanaka, M. Hoshina, *Grain size effect on crack nucleation and growth in long-life fatigue of low-carbon steel*, Fatigue Mechanisms edited by J.T. Fong, ASTM STP 675, 1979.

[27] E. Schmid, W. Boas, *Plasticity of crystals with special reference to metals*, F.A. Hughes &Co. Limited, London, 1950.

[28] Y.Q. Chen, S.P. Pan, M.Z. Zhou, D.Q. Yi, D.Z. Xu, Y.F. Xu, *Effects of inclusions, grain boundaries and grain orientations on the fatigue crack initiation and propagation behavior of 2524-T3 Al alloy*, Materials Scinece & Engineering A, 580: 150-158, 2013.

[29] S.S. Chakravarthula, Y. Qiao, *Fatigue crack growth in a coarse-grained iron-silicon alloy*, International Journal of Fatigue, 27: 1210-1214, 2005.

[30] L. Wei, Q. Pan, H. Huang, L. Feng, Y. Wang, *Influence of grain structure and crystallographic orientation on fatigue crack propagation behavior of 7050 alloy thick plate*, International Journal of Fatigue, doi: http://dx.doi.org/10.1016/j.ijfatigue.2014.03.009, 2014.

[31] H.F. Poulsen, *Three-Dimensional X-ray Diffraction Microscopy – Mapping Polycrystals and their Dynamics*, Springer Publisher, ISBN 978-3-540-22330-6, 2004.

[32] W. Yang, B.C. Larson, J.Z. Tischler, G.E. Ice, J.D. Budai, W. Liu, *Differential-aperture X-ray structural microscopy: a submicron-resolution three-dimensional probe of local microstructure and strain*, Micron, 35: 431-439, 2004.

[33] P. Reischig, A. King, L. Nervo, N. Vigano, Y. Guilhem, W.J. Palenstijn, K.J. Batenburg, M. Preuss, W. Ludwig, *Advances in X-ray diffraction contrast tomography: flexibility in the setup geometry and application to multiphase materials*, Journal of Applied Crystallograpy, 46: 297-311, 2013.

In: Recent Trends in Fatigue Design
Editor: Ricardo Branco
ISBN: 978-1-61668-410-5

Chapter 8

STRAIN ENERGY RELEASE RATE AND TEARING MODULUS IN STRUCTURAL MATERIALS WITH VARIED CRACK GEOMETRY

***K. V. Arun*[1*], *K. K. Jadhav*[2] *and Virupaxi Bagodi*[3]**
[1,3]Department of Mechanical Engineering, Government Engineering College, Haveri, Karnataka, India
[2]Union Public Service Commission, New Delhi, India

ABSTRACT

Cracks in structures are the highest stress concentrators and favor their rupture. Selecting a material that exhibits significant resistance to crack growth can frequently be a critical decision. The main parameters which influence on the rate of crack propagation are the strain energy release rate and the tearing modulus. This investigation aims at the evaluation of the effect of crack configuration on these parameters in the structural materials. The experimentations have been carried out on three different crack configurations, like, single edged notched bars (SENB), double edged notched bars (DENB) and circumferentially notched bars (CNB). The strain energy release rate and the tearing modulus were determined by the experimental compliance method. The results of the experimentations have shown that the rate of crack growth has a

* bdt.arun@gmail.com, +91 9844533535.

maximum dependency on the crack configuration and its depth. The rate of crack growth has shown maximum dependency on the stretched zone width, which in turn has shown dependency on the crack geometry. The experimentations were carried out on the C15 steel and 7475 aluminum. The specimen preparation and the experiments were conducted according to the ASTM standards. The micromechanism of the crack growth and the fracture has been analyzed through SEM fractography.

Keywords: Compliance, Strain Energy Release Rate, Tearing Modulus, Pre-Cracked Round Bars, SEM Fractography, Stretched Zone

INTRODUCTION

Crack initiation data is of vital importance to the component designer, but the behavior of a material *after* a crack has initiated is also a major concern in many applications. Selecting a material that exhibits significant resistance to crack growth can frequently be a critical decision. Subcritical crack growth can be observed in machine and structural components made of ductile or semi ductile structural materials. It is of great importance to understand this process properly in order to predict the onset of unstable crack growth [1]. One type of crack propagation test involves the determination of crack growth per load cycle as a function of stress-intensity factor range [2,3]. The J contour integral is extensively used in fracture mechanics, as both the energy and the stress based criteria, for determining the onset of crack growth [1]. The main parameters which influence on the rate of crack propagation are the strain energy release rate and the tearing modulus. The J integral is the equivalent of the G for the elastic-plastic case. It is the rate of energy absorbed per unit area as the crack grows; it is not however the energy release rate because the plastic energy is not recoverable as it would be in the elastic case. The area on load displacement diagram at a particular displacement represents the energy released in N-mm. The energy release rate G, can be calculated by LEFM approach. Also the energy release per unit diameter per unit crack length is represented as J, in case of EPFM approach. Also it is identified that the main parameters, which influence on the rate of crack propagation are the strain energy release rate (SERR) and the tearing modulus [2, 4, 5]. Many researchers [6-8] have carried out experimental investigations on strain energy release rate and the tearing modulus. But less amount of work has been reported on these parameters in the pre cracked round bars with different crack configurations. It is also well documented in the previous literature [3, 5] that the experimental compliance method is the best suited technique for the evaluation of these parameters.

A stretched zone is formed at the tip of the fatigue crack in a fracture toughness specimen before fracture occurs. There has been some speculation as to the significance of the stretched zone. The stretched zone observed at the fatigue precrack tip of fractured specimen, firstly noticed by Spitzig [9], is considered to be an important parameter which represents the behavior of the precrack tip of the specimen subjected to increasing load leading to fracture [10-13]. Observation of fracture surfaces is of paramount importance in fracture investigation. Such an investigation helps to identify the cause of fracture.

MATERIALS AND EXPERIMENTATION

Experimentation is carried out on the C15 steel and 7475 aluminum, which are widely used structural materials in automobiles, aerospace, ship building, nuclear power plants, machine tools and general engineering applications. The chemical composition and the properties of these materials are as given in the tables 1 and 2.

To do a Fracture Mechanics analysis, the strain energy release rate and tearing modulus are to be known for the geometry of interest. For many geometries these parameters, both theoretical and experimental have been determined [2,6]. But for solid and hollow round bars with different crack configurations, little work has been reported. In the present work, investigations have been carried out on single edge notched bars (SENB), double edge notched bars (DENB) and circumferentially notched bars (CNB). The specimens were prepared by the standard bar stocks of the respective materials and fine notches of each type are created in these specimens. For all the notch configurations, the specimens were investigated with different notch depths. The specimen preparation and the experimentations were carried out according to the ASTM standards.

Table 1. Composition and Properties of C15 Steel

Element	C	Fe	Mn	S	Ph
%	0.14 - 2	99.13-99.58	0.3-0.6	Max 0.05	Max 0.04

Yield strength = 310 MPa Ultimate strength = 375 MPa Young's Modulus = 205 GPa.

Table 2. Composition and Properties of 7475 Aluminum

Element	Al	Cr	Mg	Mn	Si	Zn	Fe
%	93-96	0.05-0.25	3.5-4.5	0.2-0.7	0.4	0.25	0.5

Yield strength = 255 MPa Ultimate strength = 295 MPa Young's Modulus = 70.3 GPa.

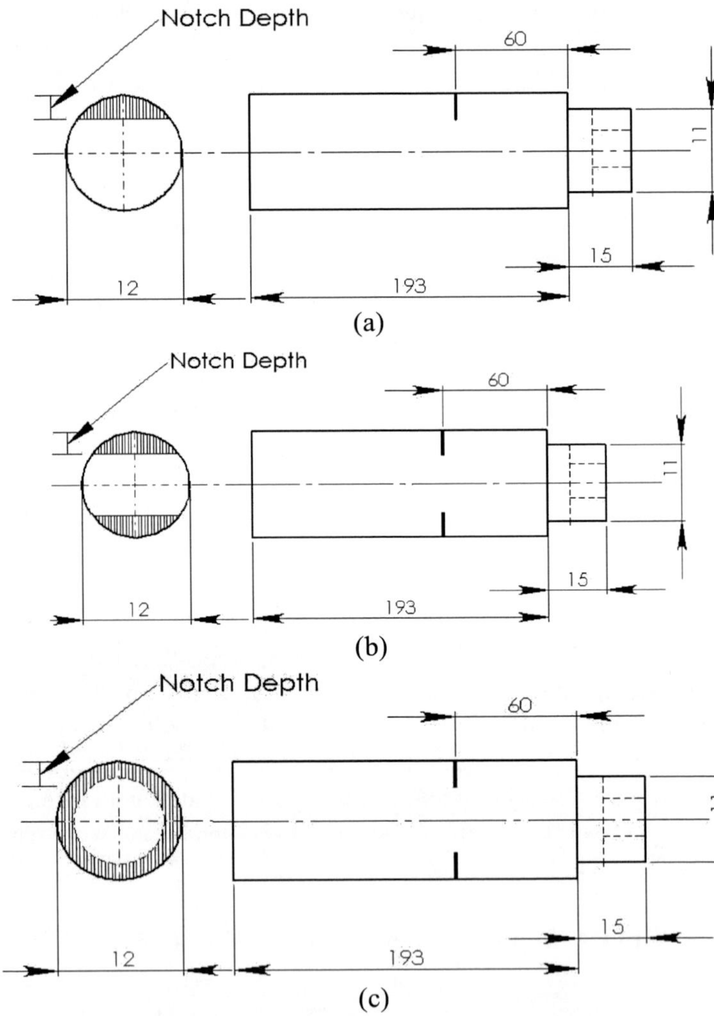

Figure 1. Notched Round Bar Specimens, a) SENB, b) DENB, c) CNB.

A literature review and the market study were performed to determine the possible ways to pre crack the specimens. It was found that the most appropriate method to generate a crack in a notched round bars, with minor eccentricity, is the rotating bending fatigue technique [14]. Thus the notched bars were pre cracked under mode-I condition in a fatigue testing machine. The initiation of crack was done on completely reversed bending fatigue loading of the specimen to a maximum nominal stress of 0.6 to 0.7 times the yield strength. The mode-I pre cracked specimens were then fractured under uniaxial tensile load on a computerized universal testing machine. The specimens used in this experimentation are as shown in the figure 1.

RESULTS AND DISCUSSIONS

Experimentations have been carried out to investigate the effect of crack geometry on the strain energy release rate and the tearing modulus. These crack tip characterizing parameters have been computed by experimental compliance technique. Also the micromechanism of crack growth is analyzed through SEM fractography. The results of which have been discussed below.

Effect of Crack Geometry on the Compliance

Form the load displacement curves, the compliance was calculated, and the curves of compliance versus crack length were plotted. Mainly the compliance curves are used as the basic tools in the determination of strain energy release rate. These curves will also give the information regarding the crack front propagation rate. The results of the experimental investigations made on the pre cracked round bars of C15 steel with all circumferential notch has been presented in the figure 2.

The failure of such structural materials is mainly because of the unstable crack propagation. The compliance is computed by the load displacement diagrams and has shown the same trend as the one reported in [15]. From the experimental investigation it is found that, as the crack length increases the compliance (i.e. inverse slope of the load, displacement curve) will also increase. To a certain depth of crack length (stable cracks) the increment in compliance is less. But at the higher crack lengths (unstable cracks), the compliance will increase drastically. For different notch configurations and for both the materials, the same curves have been plotted and the trend of the curves is found to be same.

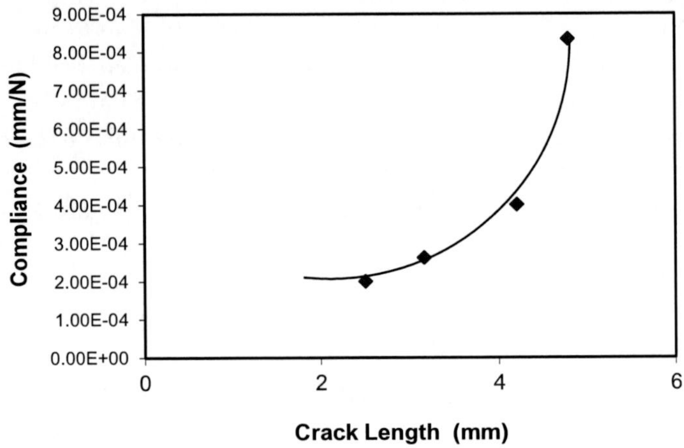

Figure 2. Variation of Compliance with Respect to the Crack Length in C15 Steel (CNB).

Effect of Crack Geometry Strain Energy Release Rate

Most of the present ductile fracture theory is expressed in terms of J integral, which perhaps is the fracture resistance of the material. The studies made by many researchers [16, 17] have shown that the value of J integral is strongly dependent on the specimen geometry. Due to insufficient thickness to maintain plane strain conditions, the formation of large plastic zones and elastic plastic behavior of the materials, the analysis of the SERR J has been made through EPFM approach. For applications, crack tip is assumed stationary, strains are assumed to increase monotonically with time during loading. Also the stress field energy density is considered to be the same by assuming reversible stress-strain behavior. Figures 3 and 4 show the trend of variation of strain energy release rate (SERR) J, with reference to the crack displacement. The curves have been shown for the same crack depth (2.44 mm) for all the three types of crack configurations.

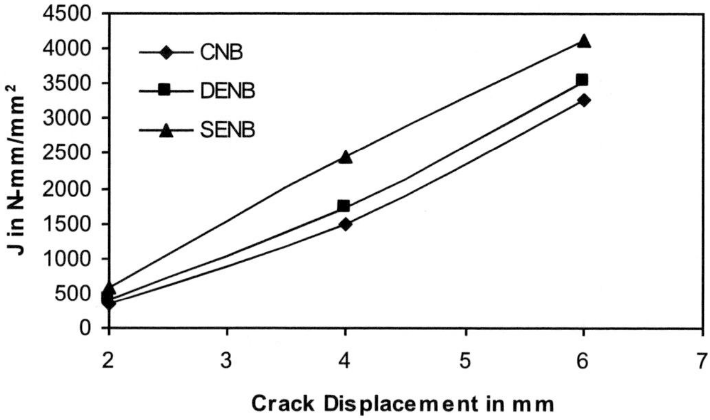

Figure 3. Variation of SERR J with respect to the Crack Displacement in C15 Steel.

It is clear that the size of the region (plastic zone) covered by the plastic flow of the material at the end of the crack tip, depends on material properties, the shape of crack and the material ahead of the crack tip. The experimental results of the variation of J for the materials under normal condition are comparable with those reported in [18,19], for circumferentially cracked round bars. It is noticed from the figures 3 and 4 that the SERR J has increased with increase in the crack extension, for all the three types of crack configurations. The experimental results have shown that the material has undergone significant plastic deformation prior to initiation of crack growth. With these material characteristics, crack growth initiation is followed by stable crack growth. The variation in the SERR has shown greatest dependency on the crack length, plastic zone size, and stretched zone width.

The same trend of variation is observed in CNB, SENB and DENB specimens. In case of the SENB the increase in the SERR with respect to the increment in crack displacement is mainly because of the increase in the plastic zone size, as the crack progresses. It is noted that, as the crack length increases, the crack width has also increased. This is mainly because of the round shape of the specimen. This increase in the crack width has increased the crack growth resistance and in turn the value of J. It is noted from the figures 4 and 5 that the maximum value of J is found in the SENB and minimum in CNB. From this it can be noticed that the stable crack growth is more in the SENB and the same is very less is CNB. A moderate behavior is observed in the DENB. From this analysis it can be very clearly noticed that circumferential cracks are most dangerous.

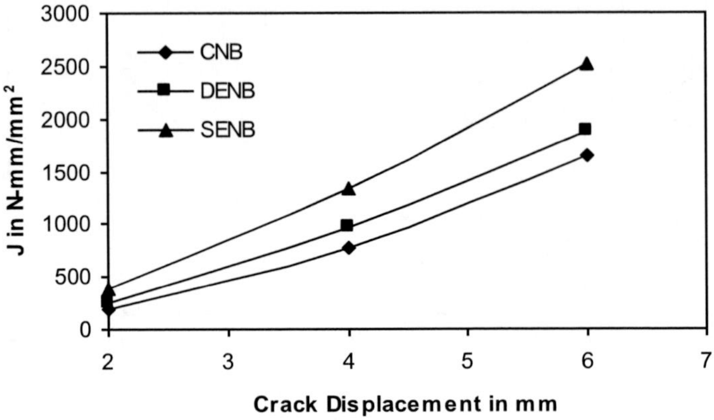

Figure 4. Variation of SERR J with Respect to the Crack Displacement in Aluminium.

Effect of Crack Geometry Tearing Modulus

The size of flaw and the coating thickness have shown maximum influence on strain energy release rate (both G and J). This is mainly affected by the size of the plastic zone ahead of the crack tip and which in turn depends on the crack size and the coating thickness. The stability analysis of the cracks in the materials loaded under tensile loading is also made through tearing modulus approach [6, 20]. Consideration in tearing modulus approach is the proportional loading of the crack tip field must be satisfied during the crack growth. The condition for the proportional loading (J controlled growth) is (dJ/da) *(B/J) >>1, where B is the remaining ligament, and the term on the left side of the inequality generally is denoted as ω. When ω is greater than 10 J, controlled growth requirements are satisfied and J/T theory is applicable. This condition is maintained in the present investigation and all the three materials have been tested under normal and coated conditions with different crack lengths.

In the present investigation the tearing modulus values were computed at three displacements, for all the materials under all conditions. It is observed that the tearing modulus has increased systematically with the increase in crack displacement. The effect of the remaining material ligament ahead of the crack tip and crack geometry on the tearing modulus, have been analyzed. Mainly the variation of the tearing modulus in coated and uncoated materials is because of the nature and size of the stretched zone formed during the crack extension. The width of the stretched zone has shown maximum influence on the tearing modulus, which in turn depends on the ductility and yield strength of the materials.

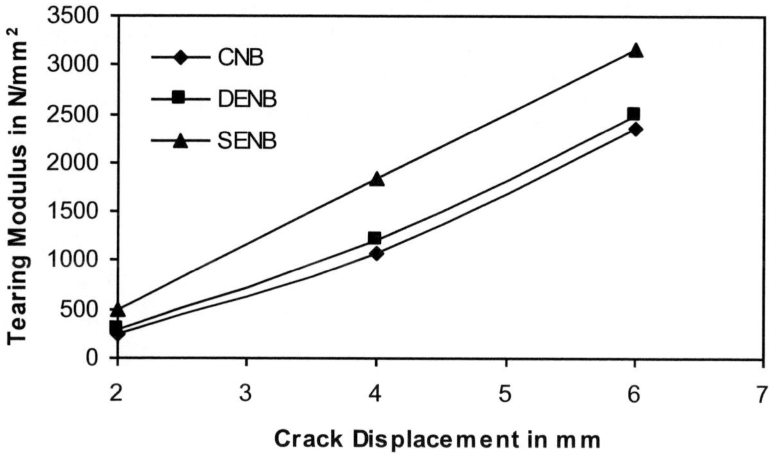

Figure 5. Variation of Tearing Modulus with Respect to the Crack Displacement in C15 Steel.

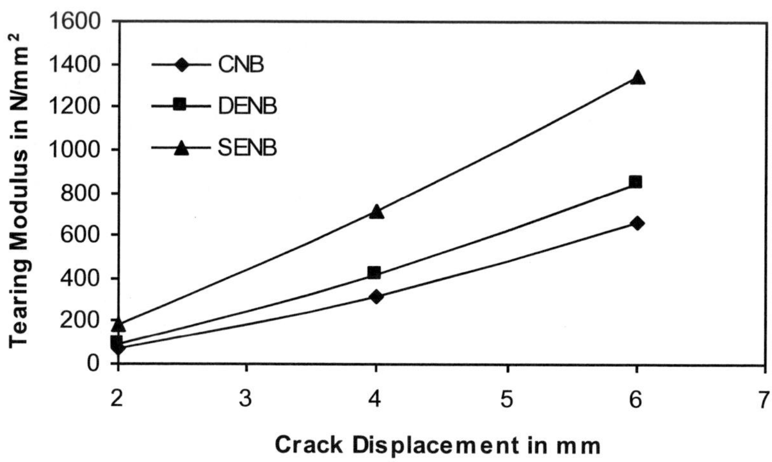

Figure 6. Variation of Tearing Modulus with Respect to the Crack Displacement in Aluminium.

Figures 5 and 6 show the variation in tearing modulus with respect to the crack displacement. From the figures it is noted that, for both the materials and for all the three crack geometries the tearing modulus has increased with increase in the crack displacement. The maximum tearing modulus is obtained for single edge cracked specimens and the least is observed in the circumferentially cracked specimens. This analysis shows that the crack growth is stable for maximum displacement of crack, in case of SENB and the crack growth will occur in an unstable manner in case of CNB. By this it can be noted that the circumferential cracks are most dangerous in nature, which may lead to the catastrophic failure of the structure.

Micromechanism of Fracture

The fractured surfaces of the samples were closely examined by means of SEM techniques, particularly in relation to the site of initiation of fatigue cracks and the different stages of their subsequent propagation. All the fractured specimens were scanned in the regions of fatigue crack, just below the fatigue crack tip and away from the crack tip along the entire fractured surface. The fractured surfaces at the boundary between fatigue crack propagation and unstable fracture were examined in a scanning electron microscope. By analyzing these scans the nature of the bonding can be predicted. The deposit (coating) seems to be uniform and has satisfactory adhesion characteristics due to the absence visible cracks at the interface.

Once the crack starts propagating under uniaxial tensile loading, it is found that for all the crack geometries the trend of variation in J and T is almost the same, but the intensity has different values (Figures 3-6). Fracture of the structural materials is usually preceded by definite microscopic processes taking place near the crack tip. The tensile loading of the cracked specimens virtually obliterates the surface marks and the mechanism that causes this relatively featureless appearance is called stretching. Fractographic observation has shown that there is a cluster of such fractured facets at the junction between the cracked and the tensile broken areas. This region is called Stretched Zone (SZ).

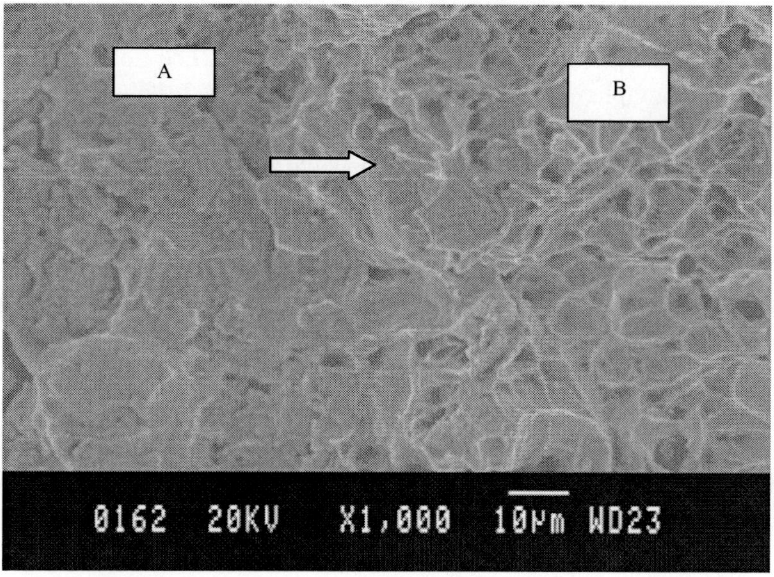

Figure 7. SEM Scan of the Stretched Zone in C15 Steel.

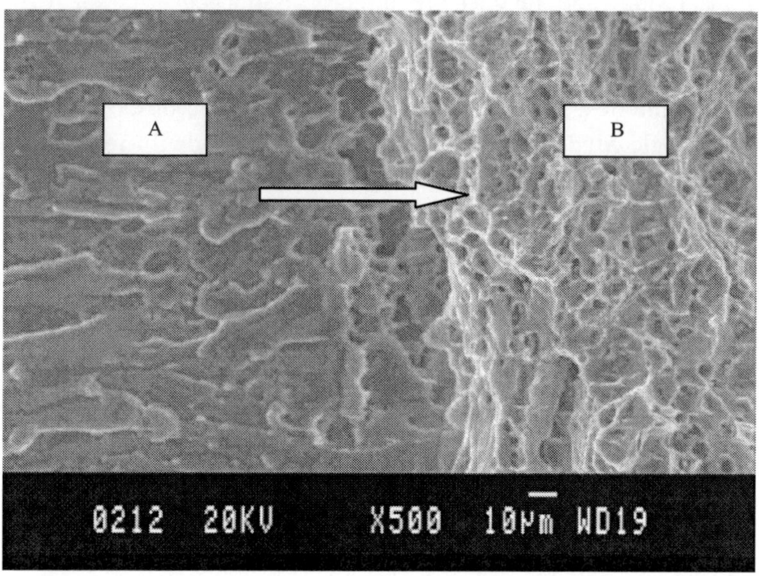

Figure 8. SEM Scan of the Stretched Zone in 7475 Aluminum.

A stretched zone is formed at the tip of the fatigue crack in a fracture toughness specimen before fracture occurs. The stretched zone is visible on electron micrographs of the transition from fatigue to ductile rupture. It consists of coarse wavy slip lines. The stretched zone occurs at the root of the fatigue pre crack, and it is postulated that the extent of stretched zone is a function of the amount of localized plastic flow at the crack tip before the onset of rapid crack propagation. Figures 7 and 8 show the formation of stretched zone between the fatigue pre crack and the final fracture regions, in C15 steel and 7475 aluminum materials. From the figures it can be seen that the stretched zone will form at the interface between the fatigue cracked and tensile broken region (between the regions A and B, as shown by the blocked boxes). The stretched zone in the circumferentially cracked bars of both the materials is shown in the figures (by blocked arrows).

It is very clear from the scans that the width of the stretched zone has varied with respect to the type of the material. The results have shown this stretched zone has appeared during the crack tip blunting under the tensile load, before the onset of rapid unstable crack propagation, which leads to the catastrophic fracture. The experimental data proved that the stretched zone dimensions (width and height) are directly associated with the SERR and tearing modulus of these materials. Based on the fractographic analysis, it is noticed that the dimensions of the stretched zone are very much affected by the crack geometries. The width of the stretched zone observed is higher in case of the single edge cracks and moderate in case of double edge cracks. But it is found to be very less in case of circumferential cracks, which in turn has proved that these cracks are most dangerous.

CONCLUSION

From the experimental and the fractographic analysis carried out on the C15 steel and aluminum, with three different crack geometries, the following conclusions were drawn:

- To certain depth of crack length the increment in compliance is less but at higher crack lengths compliance increases drastically for both the structural materials. This behavior of the compliance is because of the stretched zone formation.

- The crack geometry has shown maximum influence on these fracture SERR and tearing modulus, which determines the plastic zone ahead of the crack tip.
- The stretch zone width (SZW) has a maximum influence on the rate of crack growth and the failure of the component. The SZW mainly depends on the ductility of the material and the crack geometry.
- The experimental and fractographic analysis has shown that the rate of fatigue crack initiation and propagation is high in case of the circumferential cracks, and is slow in case of single edge cracks. Finally it can be concluded that the circumferential cracks are most dangerous in nature and may lead to catastrophic failures.

REFERENCES

[1] Neimitz, I.dzioba and J. Galkiewicz, “A Study of Stable Crack Growth Using Experimental Methods, Finite Elements And Fractography”, *Engineering Fracture Mechanics,* Vol. 71, 2004, pp.1325-1355.

[2] Bush. A.J "Experimentally determined stress intensity factors for single-edge-crack round bars loaded in bending". *Experimental Mechanics.* Vol.16. No.7. July1976, pp.249-280.

[3] Bush, A.J., “Stress Intensity Factors for Single-edge-crack Solid and Hollow Round Bars Loaded in tension,” *Journal of Testing and Evaluation JTEVA,* Vol9, No 4, July 1981, pp. 216-223.

[4] Daoud O.E.K., D.J. Cartwright, M.Carney. "Strain-Energy Release Rate for a Single-Edge-Cracked Circular Bar in Tension". *Journal of Strain Analysis* Vol-13 No.2 (1978) pp. 83-89.

[5] Hollstein, T. Schmitt and J.G. Blauel, “Numerical Analysis Of Ductile Fracture Experiments Using Single Edged Notched Tension Specimens”, *Journal of Testing and Evaluation JTEVA.* Vol.11. No.3 May-1983, pp. 176-181.

[6] Susil K Putatunda, S K, Rigsbee, J M, and Corted H T, “Effect of Size on Elastic-Plastic Fracture Toughness Parameters,” *Journal of Testing and Evaluation,* JTEVA, Vol. 13, No.3, May 1985, pp.181-190.

[7] G G Chell, “Fracture Mechanics in the Elastic-Plastic Regime”, Materials Division, *Central Electricity Research Laboratories, Surrey,* UK.

[8] L Towers and S J Garwood, *"The Geometry Dependence and Significance of Maximum Load Toughness Values",* The Welding Institute, Cambridgeshire, UK.

[9] W. A. Spitzig. A fractographic feature of plane-strain fracture in 0.BC-Ni-Cr-Mo steel. *Trans.* ASM 61,344 (1%8).

[10] J. P. Tanaka, C. A. Pamillo and J. R. Low, Jr., Fractographic analysis of the low energy fracture of aluminum alloy. *ASTM Spec. Tech.* Pi&. 463, 191 (1970).

[11] K. Schwdbe, Processes at the tip of a crack under continuously increasing loads. *3rd Int. Conj. Fmctun*, m-515. Munich, Teil-IV (1973).

[12] D. Elliot, Crack tip processes leading to fracture. *The Practical Implications ojFmctun Mechanisms*. pp. 21-27. The Inst. of Metallurgists (1973).

[13] D. Broek, Correlations between stretched zone size and fracture toughness. *Eng. Frac.* Me& 6, 173 (1974).

[14] Marc Scibetta, "*Ph.D Thesis on Contribution to the Evaluation of the Circumferentially Cracked Round Bar for Fracture Toughness Determination of Reactor Pressure Vessel Steels".* March 1999.

[15] Ashok Saxena and S J Hudak," Review and Extension of Compliance Information for Common Crack Growth Specimens", *International Journal of Fracture,* Vol.14, October 1978, pp.453-468.

[16] D T Read and R P. Reed, "Effects of Specimen Thickness on Fracture Toughness of an Aluminum Alloy", *International Journal of Fracture,* Vol.13, April 1977, pp.201-213.

[17] J G Merkle and H T Corten, "A J Integral Analysis For The Compact Specimen, Considering Axial Force As Well As Bending Effects", *Journal of Pressure Vessel Technology,* 1974.

[18] J C Devaux and G Rousselier, " An Experimental Program For The Validation Of Local Ductile Fracture Criteria Using Axisymmetrically Cracked Bars And Compact Tension Specimens", *Engineering Fracture Mechanics, Vol.* 21 , 1985, pp.273-283.

[19] T Nakamura, C F Shih and L B Freund, " Elastic- Plastic Analysis Of A Dynamically Loaded Circumferentially Notched Round Bar", *Engineering Fracture Mechanics,* Vol. 22 , 1985, pp.437-452.

[20] N G Awadalla, R L Sindelar, G A Abramczyk, *"Fracture Toughness Of Type 304 Stainless Steel",* Technical Report, Reactor Materials Program, sept.1986.

INDEX

D

E

F

G

H

I

J

K

L

M

N

O

P

W

X

Y